OrCAD® PSpice® for Windows

Volume II: Devices, Circuits, and Operational Amplifiers

Third Edition

Roy W. Goody
Mission College

Pearson Education

Upper Saddle River, New Jersey
Columbus, Ohio

Vice President and Publisher: Dave Garza
Editor in Chief: Stephen Helba
Acquisitions Editor: Scott J. Sambucci
Production Editor: Rex Davidson
Design Coordinator: Robin G. Chukes
Cover Designer: Becky Kulka
Cover Art: FPG
Production Manager: Pat Tonneman
Marketing Manager: Ben Leonard

The book was printed and bound by Victor Graphics, Inc.
The cover was printed by Victor Graphics, Inc.

OrCAD® and PSpice® are registered trademarks of Cadence Design Systems.

10 9 8 7 6 5 4 3 2 1

Pearson
Education

ISBN 0-13-015797-X

Contents

Preface

What's past is prologue.
WILLIAM SHAKESPEARE, *THE TEMPEST* (1600)

This is the second in a three-volume series on PSpice circuit simulation and covers *Devices, Circuits,* and *Operational Amplifiers.* (Volume I covers *DC and AC Circuits* and Volume III covers *Digital and Data Communications.*) If you are already familiar with Volume I, you will find few surprises in Volume II. If not, a brief overview is provided here.

The material is based on OrCAD Lite 9.2, the latest free evaluation version of the most popular simulation software on the market today: OrCAD corporation's PSpice. It is introductory in nature and is appropriate for those with little or no experience in circuit simulation. The level of difficulty is tailored to the technology student, but it offers enough "gentle" material for the technician and enough challenging material for the engineer. It covers both PSpice techniques and analog theory and applications, and the choice and sequence of material closely follows that of a conventional theory text. Since most activities can be done by either PSpice simulation or hands-on construction, it is designed to replace a conventional laboratory manual that covers devices, circuits, and operational amplifiers.

Why PSpice?

As a dedicated student or educator, you may strongly believe that circuit simulation must be a part of your classroom experience before applying your knowledge and skills on the job. It would be reasonable, therefore, to seek out the most popular circuit simulation software on the market today. It also would be beneficial to use the software package that is used by engineers and technicians on the job—complete with professional-level advanced techniques and restricted only by circuit size. Further it would be sensible to reduce your costs to zero and distribute the free software without the need to worry about licenses or copyright restrictions. The solution, then, is PSpice.

Volume II

We assume that the majority of students reading this preface have at least a passing familiarity with *OrCAD PSpice for Windows*, Volume I, covering DC and AC circuits. Volume I also covers the most fundamental PSpice techniques and processes, and there is not sufficient room in Volumes II or III to repeat this introductory material. We can, however, present short review inserts where appropriate and reprint several of the special Simulation Notes in Appendix A. Therefore, we recommend that you keep a copy of Volume I available for reference and review.

Volume II is divided into six parts: Diode Circuits, Bipolar Transistor Circuits, Field-Effect Transistor Circuits, Special Solid State Studies, Operational Amplifiers, and Special Processes. This is the same order and mix of subjects normally found in a typical devices and circuits and analog course. The special processes of Part 6 illustrate the benefits of using circuit simulation over conventional prototyping. For the most part it is not necessary to complete the first five parts before turning to and dabbling in these powerful techniques.

OrCAD's Total Solution

For designing electronic circuits, OrCAD offers a total solution package, including schematic entry, FPGA synthesis, digital, analog, mixed-signal simulation, and printed circuit board layout—everything from start to finish. All software components are fully integrated and are designed to follow an engineer's natural design flow.

This text is based exclusively on just one part of the complete package: PSpice A/D. Fortunately, PSpice A/D is precisely what we need to support a college-level technology class, for this software component simulates nearly any mix of analog and digital circuits and conveniently displays the results in graphical form. It is incredibly powerful, easy to learn, and simple to use. Quite simply, OrCAD's PSpice A/D is one of the best learning tools available.

OrCAD Lite

Fortunately, for those of us in education, OrCAD Corporation has made PSpice evaluation software available at no cost. All the activities in this book are based on OrCAD Lite version 9.2. Its only major limitation is the number of symbols and components that can be placed on the schematic. Fortunately, we can adjust easily to these limitations, and for the most part they will be completely invisible.

Third Edition

The major improvement of this third edition is the move up to version 9.2. Although many of the circuits and components are unchanged from the second edition, a greatly improved *project manager* window facilitates schematic organization, assignment of simulation type, and analysis and display of the circuit response. For this volume II, we now include operational amplifiers.

Suggestion

Although circuit simulation is the major design and development tool of the future, we recommend that the reader also receive hands-on experience by prototyping actual circuits and troubleshooting with conventional instruments.

One computer-saving approach is to divide a class into two or more groups and switch between PSpice and hands-on techniques. It is especially instructive to perform the same activity using both PSpice and hands-on techniques, and to compare the two approaches. *In this regard, most of the experimental activities outlined in this text can be performed using either PSpice or hands-on techniques.*

Further Study

If you order the complete set of manuals that comes with PSpice, you would be confronted with more than one thousand pages of data, instructions, and reference material. Clearly, all the information contained within those pages cannot be placed into this introductory text series. Instead, we have included only the most vital and commonly used features of PSpice. For a comprehensive description of all the features of PSpice, refer to the complete set of manuals from OrCAD.

Product manuals and many other useful items and features, including technical data, articles, techtips, and university support, can be obtained from OrCAD's website (www.orcad.com).

Mouse Conventions

Throughout this text, we will adopt the following mouse convention:

- **CLICKL** or **BOLD PRINT** *(click left once)* to select an item.
- **DCLICKL** *(double click left)* to perform an action.
- **CLICKR** *(click right once)* to open a menu.
- **CLICKLH** *(click left, hold down, and move mouse)* to drag a selected item. Release left button when placed.

Acknowledgments

I wish to express my sincere gratitude to production editor Rex Davidson and acquisitions editor Scott Sambucci of Prentice Hall. Under their careful guidance, the project steadily moved forward and was released on time. I also thank the reviewers of the manuscript: Ed Bertnolli, University of North Florida; John Brews, University of Arizona; and Ronald Rockland, New Jersey Institute of Technology.

Of course, OrCAD Corporation deserves special credit for making the OrCAD Lite evaluation disk available at no cost. Their foresight makes it possible for colleges and universities to teach circuit simulation at the professional level without breaking the ever-shrinking budget.

Thank you for adopting *OrCAD PSpice for Windows*; May you have good luck and success.

Roy W. Goody

How to obtain your free OrCAD Lite software

During the writing of this text the author relied on PSpice Beta version 9.2. Unfortunately, the free Lite version 9.2 CD (which the circuits of this text are designed for) was not available when this text went to the presses—but should be by the time it reaches your hands.

To obtain your free copy of PSpice Lite do any of the following:

- Go to OrCAD's website (www.orcad.com) and order the software as a CD to be delivered by mail.

- Download the software directly from OrCAD's website.

- Phone OrCAD sales at 1-888-671-9500.

- Instructors and professors can obtain a free copy of demo 9.2 by simply ordering the *PSpice for Windows Instructor's Guide* from Prentice Hall (www.prenticehall.com) or by calling your local representative.

Part 1

Diode Circuits

Part 1 begins our study of active devices with the diode. We cover the typical small-signal diode, the zener diode, and several common applications.

Because we assume that the major PSpice techniques of Volume I are second nature, we generally limit our PSpice instructions to reviews and reminders.

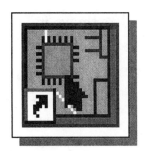

The Diode

Switching Speed

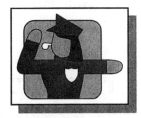

Objectives

- *To plot diode curves*
- *To examine and modify a diode's model*
- *To determine a diode's switching speed*

Discussion

As shown in Figure 1.1, a diode is an active device with a single PN junction.

Using the waterfall analogy, electrons in the high-energy conduction band can fall easily to the holes in the low-energy valence band when passing from the N to the P region (forward bias voltage). However, electrons cannot easily flow uphill from the P to the N region (reverse bias voltage). Therefore, the PN junction offers a low resistance in the forward direction and a high resistance in the reverse direction.

Note that the *schematic symbol* arrow points in the forward bias direction for *conventional* current flow (Figure 1.1).

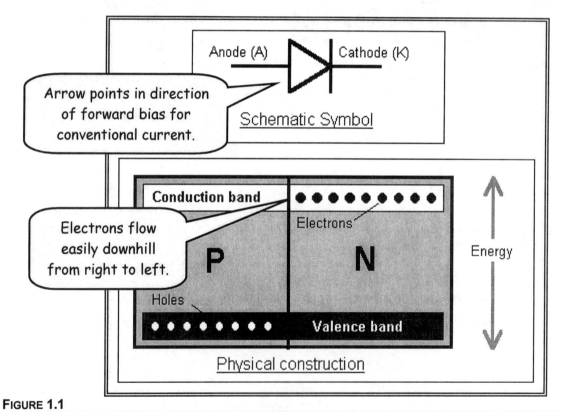

FIGURE 1.1

The PN junction

If enough voltage is applied in the reverse direction, a conventional diode will break down, and chain-reaction ionization (avalanching) will destroy the diode. A *zener diode*, on the other hand, avoids avalanche and is designed to work in the reverse breakdown region as a *voltage source* (or *voltage regulator*). Zener diodes are covered in Chapter 2.

The best way to display the characteristics of a diode is through the *diode curve*, in which diode voltage and current are plotted on the *X*- and *Y*-axes. In this chapter we draw diode curves for the 1N4148 small signal diode.

Device Models

As we learned in Volume I (Chapter 18), all PSpice devices use mathematical *models* and model *parameters* to determine their characteristics. In the case of the passive devices (*R*, *C*, and *L*), the model parameters are simple and few in number.

In contrast, active devices such as diodes have more complex models and many accessible parameters, and they give us the ability to modify parameters as needed. For the 1N4148 diode of Figure 1.1, the model parameters are listed below. In this chapter, we will access these model parameters and modify the value of the reverse breakdown voltage (*Bv*).

Parameter	Description	Value
Is	Saturation current	0.1pA
Rs	Parasitic resistance	16Ω
CJO	PN capacitance	2pF
Tt	Transit time	12ns
Bv	Breakdown voltage	100V
ibv	Reverse knee current	0.1A

We are about to begin Simulation Practice. If not already open, bring up the PSpice program by clicking the desktop icon, or **Start**, **Programs**, **OrCAD Family Release 9.2 Lite Edition**, **Capture Lite Edition**.

Simulation Practice

Activity *SMALLSIGNAL*

Activity *SMALLSIGNAL* will analyze the characteristics of the D1N4148 small signal diode.

1. Create project *diode*. (If necessary, refer to the *Create Project Review and Summary* below.)

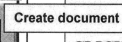

Create document

Create Project Review and Summary

CLICKL the *Create Document* toolbar button to bring up the *New Project* dialog box. Enter name *diode*, select *Analog or Mixed A/D*, and make sure the location is correct (suggest *C:\PSpice*), **OK**. When the *Create PSpice Project* dialog box opens, select (click) *Create blank project*, **OK**, to bring up the *Capture* window of Figure 1.2.

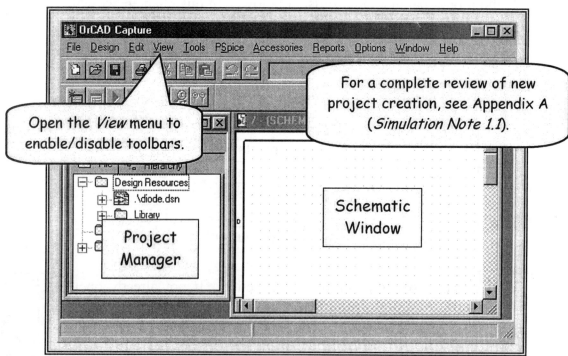

FIGURE 1.2

The *Capture* window

PSpice for Windows

2. From the *Project Manager* window, expand the project folders. Rename (**CLICKR**, **Rename**) *SCHEMATIC1* to *SMALLSIGNAL*. The default page name (*PAGE1*) can remain as is.

Add Libraries Summary

The various parts used by PSpice are stored in several libraries, all of which are recommended for this Volume II. Chances are all or part of these libraries are already listed. If not, follow the steps below:

- Click the *Place part* toolbar button to bring up the *Place Part* dialog box. **Add Library**, select (**CLICKL**) a library to add, **Open**. Repeat until all libraries are listed in the *Libraries* box.

- Click the *Place ground* toolbar button, **Add Library**, **source**, **open**.

3. Draw the test circuit of Figure 1.3. (If necessary, see Appendix A, notes 1.2 – 1.5.)

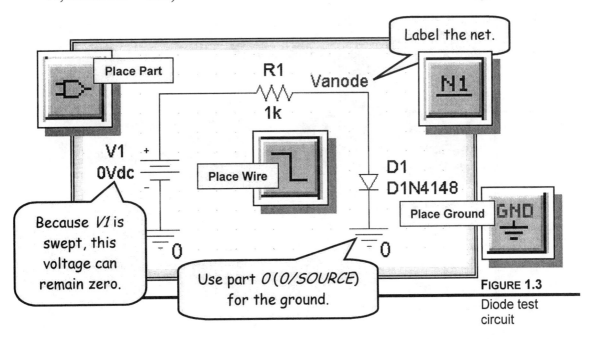

FIGURE 1.3

Diode test circuit

4. Set the simulation profile to *DC Sweep* from −110V to +10V in increments of .01V. (If necessary, see *Set Simulation Profile Summary* below.)

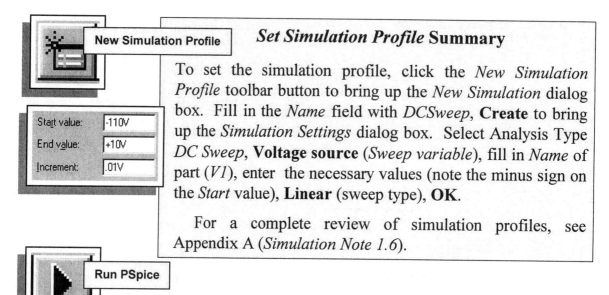

New Simulation Profile

Start value: -110V
End value: +10V
Increment: .01V

Set Simulation Profile Summary

To set the simulation profile, click the *New Simulation Profile* toolbar button to bring up the *New Simulation* dialog box. Fill in the *Name* field with *DCSweep*, **Create** to bring up the *Simulation Settings* dialog box. Select Analysis Type *DC Sweep*, **Voltage source** (*Sweep variable*), fill in *Name* of part (*V1*), enter the necessary values (note the minus sign on the *Start* value), **Linear** (sweep type), **OK**.

For a complete review of simulation profiles, see Appendix A (*Simulation Note 1.6*).

Run PSpice

5. Run PSpice, and generate the initial Probe graph of Figure 1.4.

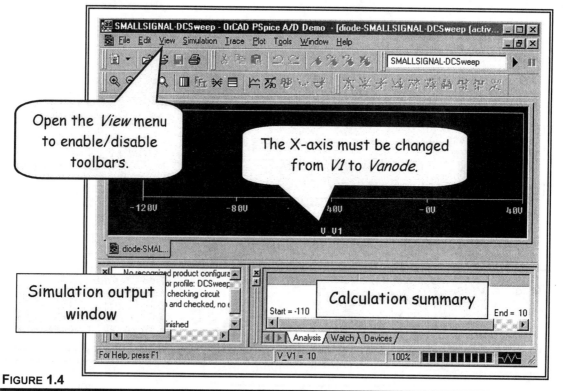

FIGURE 1.4

Initial Probe graph

PSpice for Windows

6. This time, looking at Figure 1.4, the default X-axis variable (*V_V1*) is not what we need—instead, the X-axis must be the diode's anode-to-cathode voltage (*V(Vanode)*). Review *Simulation Note 1.1* and make the change.

Simulation Note 1.1
How do I change the X-axis variable?

To change the X-axis to any available variable: **Plot**, **Axis Settings**, *X-Axis* tab, **Axis Variable**, **CLICKL** on the desired trace variable (such as *V(Vanode)*), **OK**, **OK**. (Be aware that, if appropriate, the X-axis range will automatically be adjusted to match the data base.)

For a complete discussion of axis settings, see Appendix B.

---

7. Plot diode current (*I(D1)*) on your graph and generate the curve of Figure 1.5. (See *Add Trace Review and Summary* below.)

Add Trace **Review and Summary**

To add a trace to the graph, we have the option of markers or direct trace addition.

- To use markers, click the desired marker toolbar button (from *Capture*), drag to desired circuit location, **CLICKL, CLICKR, End Mode**.

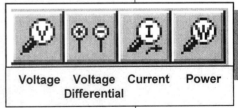

Voltage   Voltage   Current   Power
          Differential

- To use traces, click the *Add Trace* toolbar button (from *Probe*) to bring up the *Add Trace* dialog box. Click left on the desired sequence of *Output Variables* and *Functions* to build a trace expression (or enter it directly in the *Trace Expression* box), **OK**.

  For a complete review of markers and traces, see Appendix A, Volume I (*Simulation Notes 5.1 and 6.2*).

Add Trace

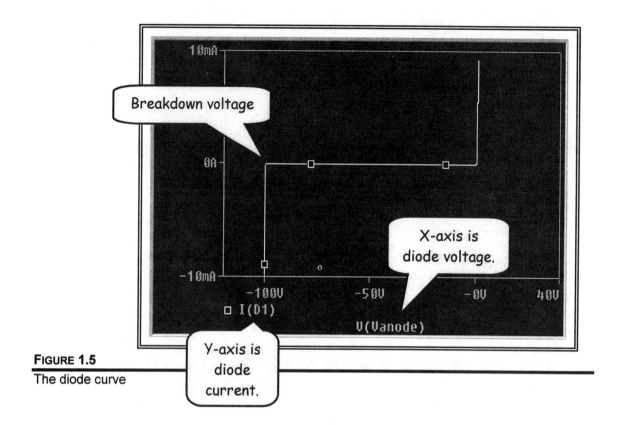

FIGURE 1.5

The diode curve

8. Based on the results of Figure 1.5, use the cursor to determine the following: (If necessary, see *Add Cursor Review and Summary* on the next page.)

    a. *Breakdown voltage* (V_{RSM})

    > V_{RSM} is also known as the *non-repetitive peak reverse voltage* and is approximately $-100V$ for the D1N4148 small signal diode.
    >
    > We will measure V_{RSM} at a typical test current of approximately $-75\mu A \pm 50\mu A$.

    V_{RSM} (at $\approx -75\mu A \pm 50\mu A$) = _____

    b. *Maximum reverse current* (I_R)   (We will measure I_R at a typical test voltage of $-90V$):

    I_R (at $\approx -90V$) = _____

***Add Cursor* Review and Summary**

To activate the two cursors (*A1* and *A2*), click the *Toggle Cursor* toolbar button. To associate a cursor with a waveform, **CLICKL** (for cursor *A1*) or **CLICKR** (for cursor *A2*) on the corresponding legend symbol. To position the cursor, **CLICKL/CLICKR** on the graph, and fine tune with the arrow keys (or Shift+arrow keys).

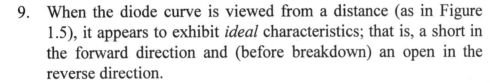

Toggle Cursor

For a complete review of cursors, see Appendix A (*Simulation Note 7.4*).

9. When the diode curve is viewed from a distance (as in Figure 1.5), it appears to exhibit *ideal* characteristics; that is, a short in the forward direction and (before breakdown) an open in the reverse direction.

   To bring out the details in the critical forward direction, zoom in on the curve by adjusting the X-axis range as shown in Figure 1.6.   (**Plot**, **Axis Settings**, *X-axis* tab, **User Defined**, fill in 0V to +1V, **OK**)

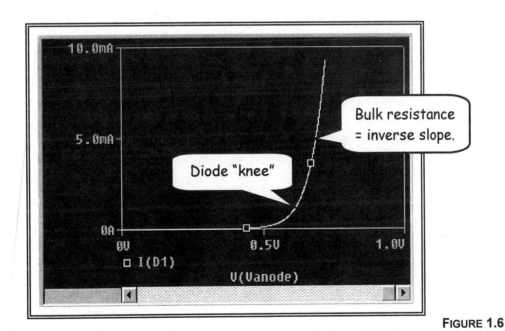

FIGURE 1.6

Forward bias details

10. From your graph, determine *approximate* values for the following (Hint: Use the cursors):

    a. The knee voltage (at 1mA): _____

    b. The forward (bulk) AC resistance of the diode at 1mA: _____

    > *Hint*: Place cursor *A1* just below 1mA and *A2* just above 1mA, and determine 1/slope as $\Delta V/\Delta I$, **or** use the "d" (differentiate) operator to plot *1/d(I(D1))* on a second Y-axis from .6V to 1V.

    c. The forward AC resistance of the diode at 5mA: _____

    d. Does the forward resistance go down as the current goes up?

    Yes   No

Model Parameters

11. Referring to *Simulation Note 1.2*, view the model parameters of the 1N4148 and change the reverse breakdown voltage (*Bv*) from 100 to 150.

Simulation Note 1.2
How do I change model parameters?

To change a model parameter: Select the component (**CLICKL**), **Edit**, **PSpice Model**, to bring up the *Model Editor* dialog box of Figure 1.7. Make the desired changes (such as *Bv=150*) and exit the window, **Yes** (save changes), **Save** (such as to file name *DIODE*).

(Saving the model to local file DIODE protects the original global model file.)

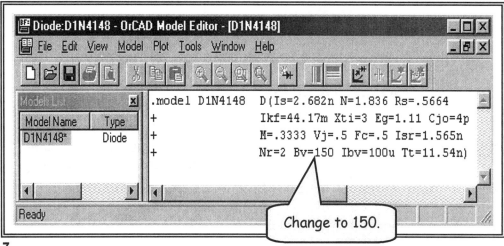

FIGURE 1.7

Diode model
parameters

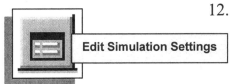

Edit Simulation Settings

12. Reset the *DCSweep* from −160 to +10V (click the *Edit Simulation Settings* toolbar button), rerun PSpice, and generate the plot of Figure 1.8. Does the breakdown now occur at −150V?

Yes   No

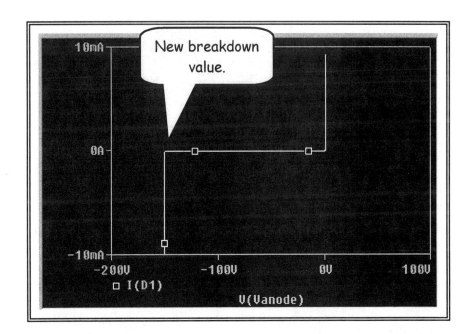

FIGURE 1.8

New breakdown
voltage.

13. By again following *Simulation Note 1.2*, return parameter *Bv* to its original default value of 100V.

Advanced Activities

14. In digital circuits, diode switching speed is very important. Create a new schematic (SWITCHSPEED), draw the pulsed circuit of Figure 1.9, select the *Transient* mode, and generate the curves of Figure 1.10.

   a. What causes the approximately 8ns of *reverse delay* in *Vanode* when V1 switches from forward to reverse bias?

   b. What causes the slight delay (≈3nS) when *V1* switches from reverse to forward bias?

   c. Based on the waveform results, what is the approximate typical (average) value of the diode's reverse-biased capacitance? (*Hint*: $I = C \, dV/dt$.)

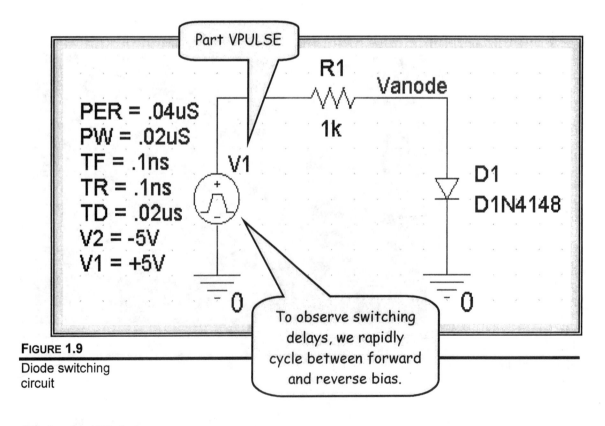

FIGURE 1.9

Diode switching circuit

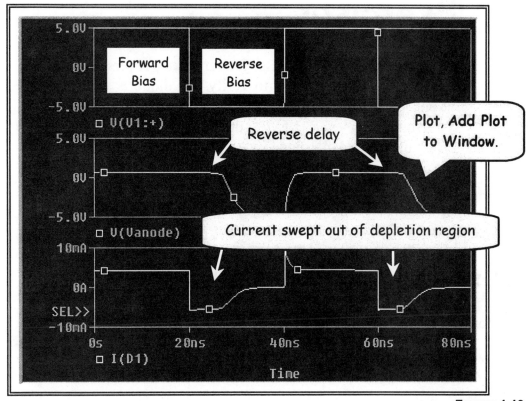

FIGURE 1.10

Switching
waveforms

15. To show how complicated component modeling can be, below is the equation for just the forward current for our "simple" 1N4148 small signal diode. (Note that many of the parameters are listed in the discussion.)

$$Ifwd = IS(e^{Vd/NVt} - 1) + ISR(e^{Vd/NRVt} - 1)((1 - Vd / VJ))^2 + 0.005)^{M/2}$$

a. Change IS (the saturation current) from 2.682n to 10n and summarize below the changes in the forward diode curve.

b. Change the *transit time* to 50ns and note the changes.

c. Change other parameters and note the waveform changes.

Exercises

1. By measuring a diode's forward and reverse resistance, an ohmmeter is often used to verify if a diode is working properly and to determine its polarity. With the assumption that a typical ohmmeter uses a 1.5V source, use this technique to test a simulated 1N4148 diode.

2. Test the switching characteristics of the D1N914 small signal diode, and compare it to the D1N4148 of Figure 1.10.

Questions and Problems

1. The forward-biased AC resistance of a diode is lowest at

   a. low currents
   b. high currents

2. Describe a quick and simple method for checking a diode using an ohmmeter.

3. What is the difference between a small-signal diode and an *LED*?

4. Based on the results of step 8b, how long would it take 1 coulomb of charge to pass through a reverse-biased diode?

5. Referring to Figure 1.11, why is the diode circuit on the left called an OR gate and the circuit on the right an AND gate?

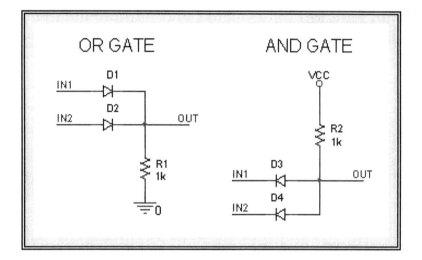

FIGURE 1.11

Diode OR and
AND gates

2

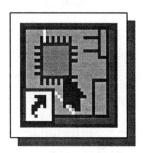

The Zener Diode

Voltage Regulation

Objectives

- *To determine zener diode characteristics*
- *To design a zener diode voltage regulator*
- *To determine how temperature changes zener characteristics*

Discussion

A voltage source is a device in which the voltage is independent of the current.   Reviewing Figure 2.1 (reproduced from Chapter 1), it appears that the breakdown region of a 1N4148 small signal diode is just such a region.

Unfortunately, as soon as the diode breaks down, the avalanche effect usually results in destruction of the diode, as ionized charges break down other atoms in a chain reaction.

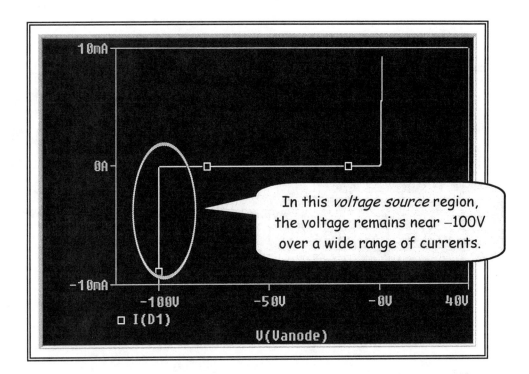

FIGURE 2.1

The diode curve

A *zener* diode, on the other hand, is designed to break down and to avoid avalanche.   Therefore, it is perfectly safe to operate in the breakdown region.   The most important application of the zener diode is its role as a voltage regulator, a device that holds the output voltage constant over a wide range of currents.

Zener diodes are available that break down over a wide range of voltages. The D1N750 diode included in the evaluation library breaks down at 4.7V. For this reason it is widely used in digital circuit power supplies.

Simulation Practice

Activity *CURVE*

Activity CURVE plots a characteristic curve for a 1N750 zener diode.

1. Create project *zener*, with schematic *CURVE* and page *PAGE1*. (Be sure to add libraries *eval* and *breakout*.)

2. Draw the zener test circuit of Figure 2.2.

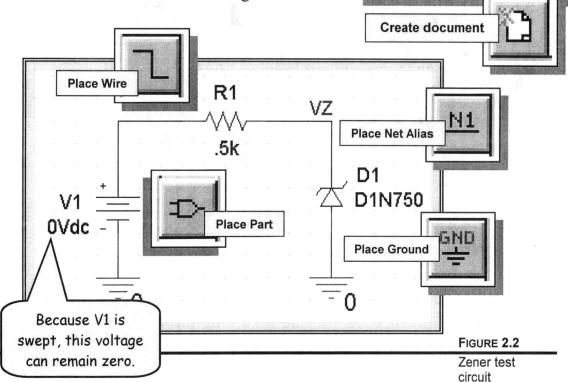

FIGURE 2.2

Zener test circuit

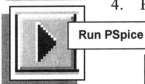
New Simulation Profile

3. Name and set the simulation profile for a linear *DCSweep* of *V1* from 0V to +20V in increments of .1V.

Run PSpice

4. Run PSpice and generate the characteristic curve of Figure 2.3.

Reminder: To change the X-axis variable: **Plot, Axis Settings, X-axis** tab (default), **Axis Variable**, enter trace expression –*V(VZ)*, **OK, OK**.

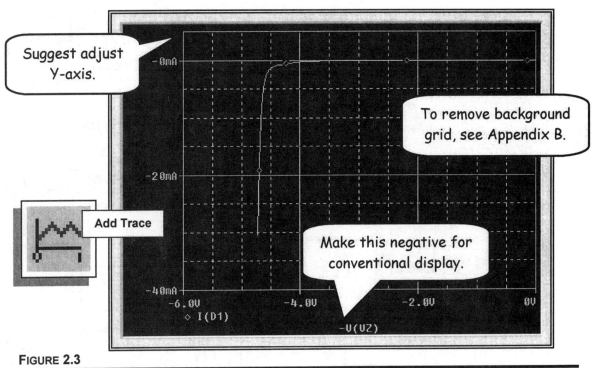

Suggest adjust Y-axis.

To remove background grid, see Appendix B.

Add Trace

Make this negative for conventional display.

FIGURE 2.3
Zener characteristic curve

5. Using a cursor, determine each of the following:

Toggle Cursor

a. V_Z (at I_Z = 10mA) = _____ V

b. V_Z (at I_Z = 20mA) = _____ V

c. V_Z (at I_Z = 30mA) = _____ V

6. Based on the results of step 5, is the zener in the breakdown region a nearly perfect *voltage source*? (Is the voltage approximately independent of the current?)

                    Yes   No

7. For the 1N750, zener characteristics are usually measured at I_Z = 20mA.

    a. Use the cursor to determine accurately the zener voltage at I_Z = 20mA.   Compare your answer with the 1N750 specification sheet in Appendix C.

        V_Z (PSpice) = _____ V_Z (spec sheet) = _____

    b. Use the cursor to determine accurately the zener impedance at I_Z = 20mA. Compare your answer with the specification sheet in Appendix C. (*Suggestion*: Choose test points slightly above and below 20mA and use the formula $Z = \Delta V/\Delta I$.)

        Z (PSpice) = _____ Z (spec sheet) = _____

Temperature Effects

8. All active devices (such as diodes) are sensitive to temperature. To see how temperature can be changed, review *Simulation Note 2.1* and change the temperature of the zener diode test circuit from the default 27°C (80.6°F) to 60°C (140°F).

Simulation Note 2.1 | Edit Simulation Settings |

How do I change the ambient temperature?

To change the circuit temperature from the default 27°C (80.6°F): Click the *Edit Simulation Settings* toolbar button, **Options** tab, change the *Default nominal temperature* to the desired value, **OK**.

---

9. Generate a new diode curve for 60°C and measure the zener voltage at 20mA. Using the data from Step 7a, as well as the new measured value of this step, fill in the table below.

VZ at 20mA (27°C)	VZ at 20mA (60°)

10. Based on the data of Step 9, by what percent did the zener voltage change due to the increase in temperature?

% change in VZ = _____ %

DC Sweep Nesting

11. A better method of evaluating temperature effects is to generate a family of curves. Follow *Simulation Note 2.2* and set up the zener temperature as a nested variable.

Simulation Note 2.2
How do I set up the DC Sweep nested mode?

The DC Sweep analysis mode offers a special *secondary* (nested) *sweep* option that operates much like a parametric sweep.

As an example of this nested mode, we set up the nested temperature curves as follows: Click the *Edit Simulation Settings* toolbar button to bring up the *Simulation Settings* dialog box and leave the *Primary Sweep* as is (*V1*, 0 to 20V, .1V increments). Click and enable the *Secondary Sweep* option and fill in the *variable* and *type* (Sweep variable **Temperature**, −50 0 +50, in increments of 25), **Apply**, **OK**.

---

12. Run PSpice, change the X-axis to −*V(VZ)*, plot the current (*I(D1)*), and generate an expanded plot similar to that of Figure 2.4. (See Appendix B for a review of axis and grid control.)

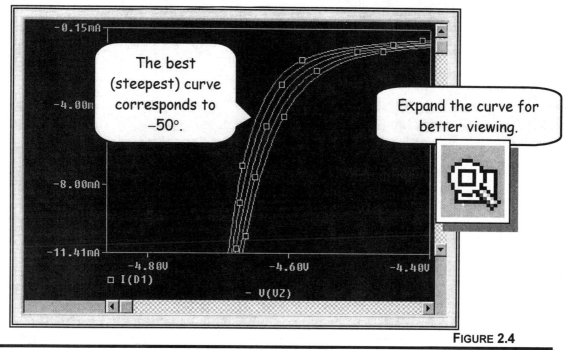

FIGURE 2.4

A *DC Sweep* nested plot

13. Why do you think the "best" (steepest) curve occurs at −50°, rather than +50°?

Advanced Activities

14. Using the "/" (divide) and "d" (differentiation) operators, add a plot of zener impedance to the zener curve of Figure 2.3 and generate the expanded plot of Figure 2.5. What is the zener impedance at −4.7V?   Do the results agree with step 7b?

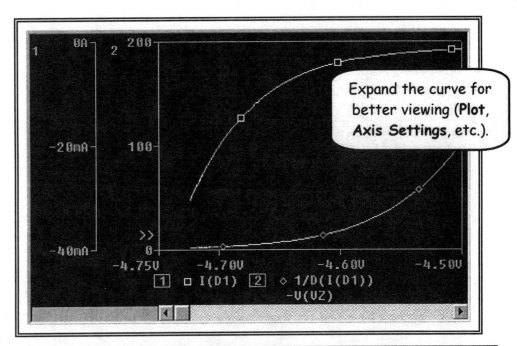

FIGURE 2.5

Adding zener
impedance

15. By way of its model parameters, change the reverse breakdown
voltage of the 1N750 zener diode from 4.7V to 10V and test
the results. At what reverse current is the zener voltage 10V?

Exercises

1. Referring to the process summary below if necessary, test the
voltage regulation characteristics of a zener diode under a
varying load. (At what value of *R2* does the zener voltage come
out of regulation and drop by 1% to 4.653?)

Process Summary for Zener Load Curve

Create schematic *LOAD* and draw the schematic of Figure 2.6. Set
the simulation profile to a primary *DC Sweep* of Global Variable
RVAL from 1Ω to 1kΩ with an increment of 10Ω. Generate the
curve of Figure 2.7.

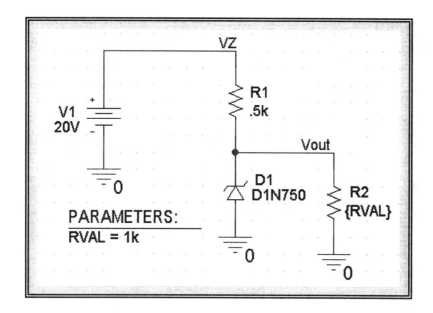

FIGURE 2.6

Zener load
test circuit

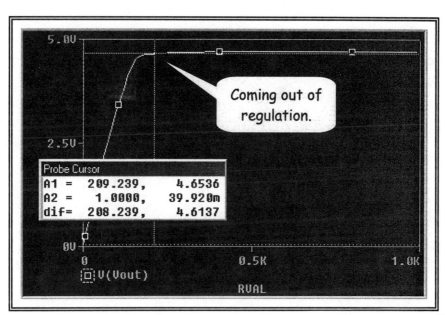

FIGURE 2.7

The Zener coming
out of regulation

2. By varying the value of *R1* (Figure 2.6), design a zener
   circuit in which the zener voltage is exactly 4.7V with a load
   of exactly 400Ω. (Assign *R1* {RVAL} and *R2* 400Ω.)

Questions and Problems

1.   A zener diode normally operates

     a. in forward bias
     b. in reverse bias

2.   What is a *voltage source*?

3.   In what region does the zener act as a voltage source?

4.   Explain the results of step 9. (Why does the zener voltage go down as the temperature goes up?)   (*Hint*: Do electrons tend to "clump up" at high or low temperatures?)

5.   Referring to Figure 2.7, approximately what is the smallest load value that will maintain zener regulation within 10%?

6.   What would the zener impedance be of an ideal zener diode?

7.   Referring to Figure 2.4, what are the primary and secondary variables?

8.   How do we remove the background grid from the probe graphs?

3

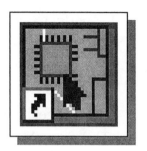

The Power Supply

AC to DC

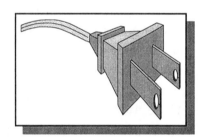

Objectives

- *To design a power supply*
- *To measure output ripple and regulation efficiency*

Discussion

A "perfect" power supply provides a constant desired voltage, regardless of the value of the load. A "real" power supply, on the other hand, has a ripple and generates an output voltage that varies with the load.

Our goal in this chapter is to design a power supply that approaches perfection, but without exceeding reasonable size, cost, and complexity limits. Our design philosophy will be to start simple and to add components and subcircuits gradually.

A Real Voltage Source

For safety reasons, it is common practice in the laboratory to use a signal generator to simulate the output from a wall socket and step-down transformer. The problem is, such laboratory voltage sources usually have a significant output impedance value. However, the voltage sources used by PSpice are "perfect" and have zero output impedance.

To simulate a laboratory voltage source using PSpice, we have the option of adding a resistor in series with the voltage source, as shown in Figure 3.1. Commonly found values of *Zout* are 50Ω and 600Ω.

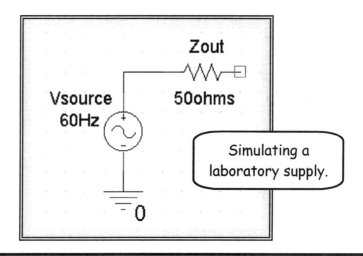

FIGURE 3.1

Simulating *Zout* of
a power source

Therefore, in this chapter the AC voltage source includes a 50Ω *Zout* resistor. In future chapters this resistor is optional but should be added whenever a PSpice simulation is to be directly compared with the same circuit built and tested in the laboratory.

Simulation Practice

Activity *GENERATOR*

Activity *GENERATOR* uses a step-by-step approach to build a three-stage power supply.

1. Create project *powersupply* with schematic *GENERATOR*.

2. Draw the initial design of Figure 3.2, which is known as a *half-wave rectifier*. (*Suggestion*: Since we will be displaying *Vout* over and over again, set a permanent voltage marker as shown.)

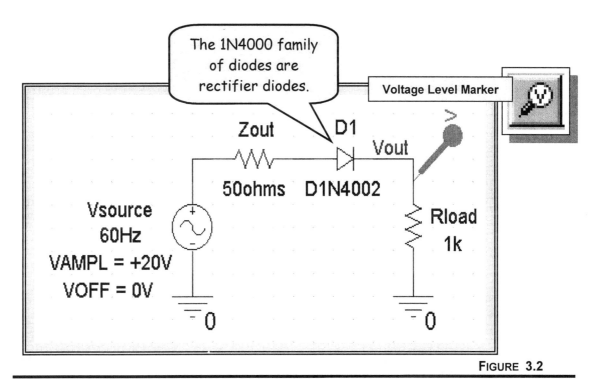

FIGURE 3.2

Half-wave rectifier

3. Set the simulation profile for *Transient* analysis from 0s to 83ms, with a step ceiling of 83μs. (*Note*: 5 × 1/60 = 83ms.)

4. Using PSpice generate the output waveform of Figure 3.3.

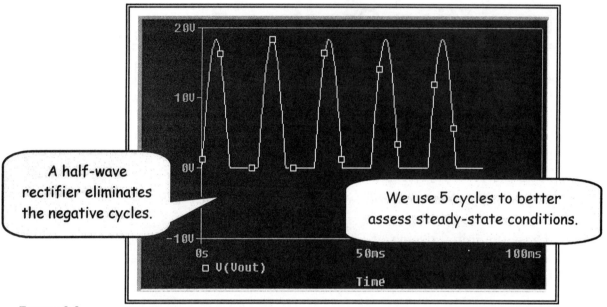

FIGURE 3.3

Half-wave rectifier
waveforms

Peak Rectifier

5. Figure 3.3 clearly tells us that we are a long way from our goal. Let's improve the circuit in two ways:

■ Use a *full-wave rectifier* so both the positive and negative input cycles will power the output.

■ Add a *filter capacitor* to smooth out the waveform.

Make these changes and create the circuit of Figure 3.4, which is known as a *full-wave peak rectifier*. The circuit acts like a leaky bucket: Each wave of voltage fills up the capacitor with charge which then leaks out through *Rload*. Due to the blocking diodes, the charge cannot return to the source.

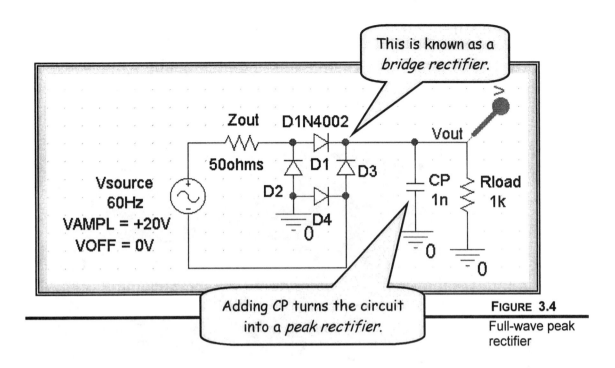

Zout   D1N4002

50ohms

Vsource
60Hz
VAMPL = +20V
VOFF = 0V

D1   D3
D2
D4
0

Vout

CP
1n

Rload
1k

0   0

Adding CP turns the circuit
into a *peak rectifier.*

FIGURE 3.4

Full-wave peak rectifier

6. To complete the design, what value will we choose for CP? (The default 1nF is much too small.) A highly accurate answer would require some advanced mathematics. Fortunately, we are only after an estimate and can make some assumptions:

- A 1V output ripple is acceptable.
- *Vout* (average) $\cong$ 16V (after Z_{OUT} and two diode drops).
- Each capacitive discharge lasts for a worst-case maximum time of 1/120 second.

Using these assumptions on the differential form of $Q = CV$ ($I = C\Delta V/\Delta t$), solve the equation below and determine the approximate value of C:

$$I \text{ (discharge current)} = C\frac{\Delta V}{\Delta t} = \frac{Vout}{Rload}$$

ΔV	=	ripple voltage	=	1V
Δt	=	maximum discharge time	=	1/120 sec
RL	=	load	=	1kΩ
Vout	=	output voltage	=	+16V

Solving for C yields approximately _____

7. Round off the value of C determined in step 6 and assign it to CP. Generate the output waveform of Figure 3.5. (*Hint*: CP ≈ 133μF.)

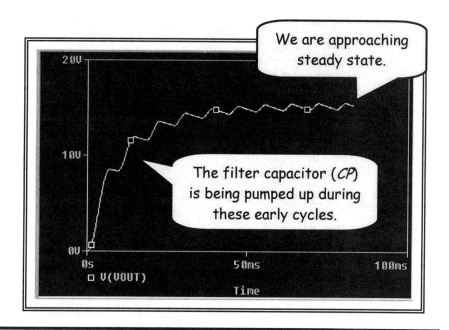

FIGURE 3.5

Peak rectifier
waveform

8. Determine the ripple voltage at the far right of the curve as it approaches steady-state conditions. (Suggestion: Position cursors at an adjacent peak and trough and report half the difference.)

    V_{RIPPLE} = _____

    Is the ripple less than 1V, as predicted by our previous calculations?

    Yes   No

Low-Pass Filter

9. We are still not satisfied with the output. The ripple voltage is too large—yet we don't wish to increase the size of the expensive and bulky *filter capacitor* (CP).

The solution is to add a *low-pass filter*, which passes the DC voltage and shorts the AC ripple to ground. Make the necessary changes and create the circuit of Figure 3.6.

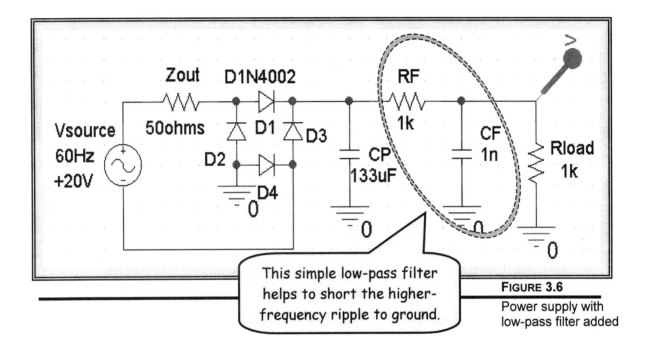

This simple low-pass filter helps to short the higher-frequency ripple to ground.

FIGURE 3.6
Power supply with low-pass filter added

10. Our next task is to change the default values of *RF* and *CF* to their proper magnitude. Again, advanced mathematics would be required for a highly accurate answer. As before, estimated values will give satisfactory results. Using the following design guidelines, determine the required values of *RF* and *CF*:

- *RF* cannot be too large compared to the load (*Rload*) because it would drop too much voltage. We arbitrarily choose 200Ω

- *CF* must have a reactance (X_C) at the ripple frequency of 120Hz that is small when compared to *RF*. We arbitrarily choose a 10-to-1 ratio, giving $X_C = 20\Omega$. Therefore:

$$\frac{1}{2\pi fC} = 20\Omega, \text{ where f(ripple frequency)} = 120\text{Hz}$$

CF = approximately _____

11. Assign to *RF* and *CF* the values determined in step 10, and generate the new (filtered) V_{OUT} of Figure 3.7. (*Hint*: $CF \approx 66\mu F$.)

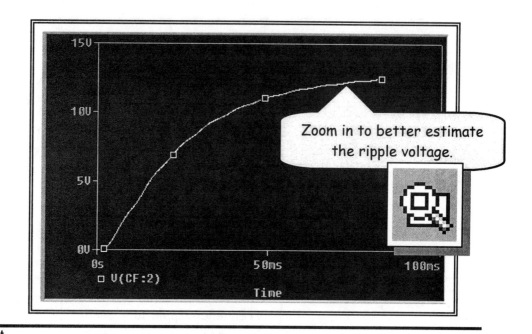

FIGURE 3.7

Filtered output
waveform

12. Determine the approximate ripple voltage at the far right of the curve as it approaches steady-state conditions. (Be sure to factor out as best you can the background slope of the curve.)

$V_{RIPPLE} \cong$ _____

Has the ripple been reduced by about 90%?

Yes   No

Voltage Regulator

13. The ripple is now quite low, but the output voltage is nowhere near our 4.7V design goal. Furthermore, the output is not regulated because *Vout* varies considerably as the load (*Rload*) changes.

To solve both problems, add the zener regulator circuit developed in Chapter 2 to create the final design of Figure 3.8. As an added bonus, we will find that the ripple voltage is further reduced.

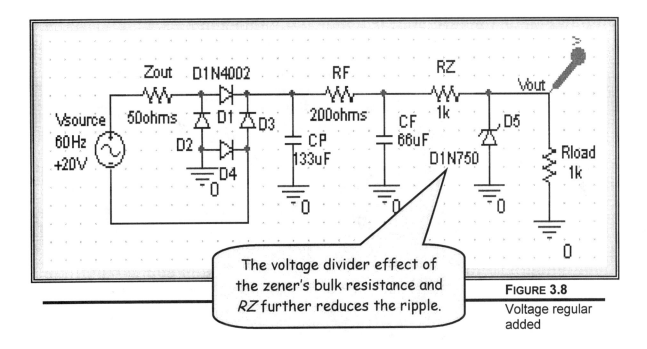

The voltage divider effect of the zener's bulk resistance and *RZ* further reduces the ripple.

FIGURE 3.8
Voltage regular added

14. To determine *RZ*, we find the desired voltage and current and use Ohm's law:

   ▪ *The current through RZ*: From Chapter 2, the approximate value of current that gives a zener voltage of 4.7V is 20mA. The 1kΩ load adds about 5mA, for a total of 25mA through RZ.

   ▪ *The voltage across RZ*: From the original 20V, we subtract 5V for the R$_{ZOUT}$/diode drop, 5V for the 25mA flowing through the 200Ω filter resistor, and 4.7V across the zener diode. This leaves approximately 5V across RZ. Therefore:

$$RZ \cong 5V/25mA \cong \text{_____} \ \Omega$$

15. Set the value of *RZ* according to step 14   (*RZ* ≈ 200Ω.), and generate the final output curve of Figure 3.9. Determine the following by using steady-state values to the far right of the curve:

    Vout = _____   Vripple = _____

    Is the output voltage near 4.7, and is the ripple very small?

    Yes   No

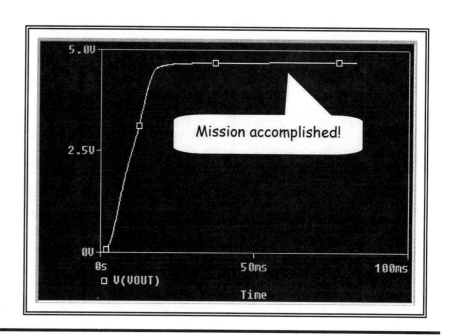

FIGURE 3.9

Regulated output

Surge Current

16. Open a second Y-axis (**Plot, Add Y-Axis**) on the graph of Figure 3.9, and add a graph of current through rectifier diodes *DR1* or *DR2* (Figure 3.10). As shown, the highest current occurs during the initial (surge) cycle (when *CP* and *CF* are empty).

    Record this *maximum surge current* (I$_{SURGE}$) below and compare with the specification sheet value (*nonrepetitive peak surge current*).

    I$_{SURGE}$ (PSpice) = _____   I$_{SURGE}$ (SPEC) = __50mA__

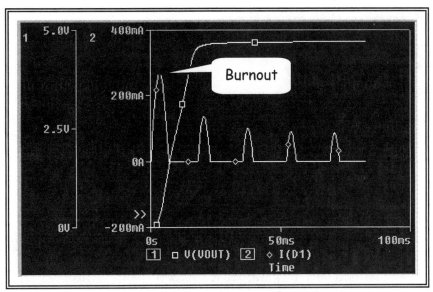

FIGURE 3.10

Diode current
added

17. Is the surge current above the maximum allowed value of 50mA (especially the first cycle)?

    Yes   No

18. To lower the surge current, add a small *surge resistor* (*RS*) to your circuit (between the output of the bridge rectifier and *CP*). Determine its value by using the worst case equation below:

    VS (peak) / I (nonrepetitive peak) = 20V / 50mA = _____

    Rerun the circuit and redisplay the diode current. (*Note*: Increase the transient time from 83ms to 150ms to reach steady state output.)

    a. Is the surge current now below the spec sheet value?

        Yes   No

    b. Is the first cycle still the worst-case surge current?

        Yes   No

Advanced Activities

Regulation Test

19.   To test the voltage regulation characteristics of our final design, generate the family of curves of Figure 3.11 by adding a nested sweep of the load value (*Rload*).   (If necessary, review the Process Summary below.)

Process Summary for Regulation Test

The main sweep variable is time, generated by a transient sweep from 0 to 1s.   The nested sweep variable is *Rload*, generated by a parametric sweep from 200Ω to 1kΩ in increments of 200Ω.   (The 400Ω surge resistor is present.) Be sure to define *Rload*'s value (*{RVAL}*) by placing and defining part *PARAMETERS*.

See Chapter 15, Volume I for a complete description of parametric analysis.

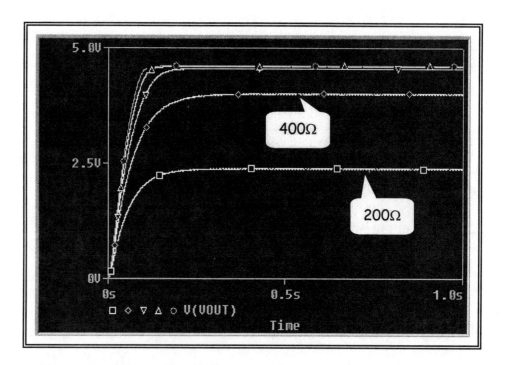

FIGURE 3.11

Regulated output
as a function of *Rload*

20. Based on your results, at what approximate value of *Rload* does the system appear to come out of regulation?

    Minimum *Rload* for regulation = _____

Transformer

21. Add a front-end transformer (from library *analog.slb*) to your power supply, as shown in Figure 3.12. In addition to its output voltage, measure the primary and secondary voltages (*vp* and *v(vs1,vs2)*) and verify the turns ratio.

    ▪ Steady-state is reached in approximately 200ms.

    ▪ $170V_{PEAK} = 120V_{RMS}$.

    ▪ To simulate a wall socket power supply, set RZ_{OUT} to a very low value.

    > *Note*: Should you encounter a convergence problem during calculation (time step goes below the minimum allowed value of 200×E−15), force the system to take larger minimum steps by setting the *Step Ceiling* in the transient analysis to an appropriate value (say, 5µs).

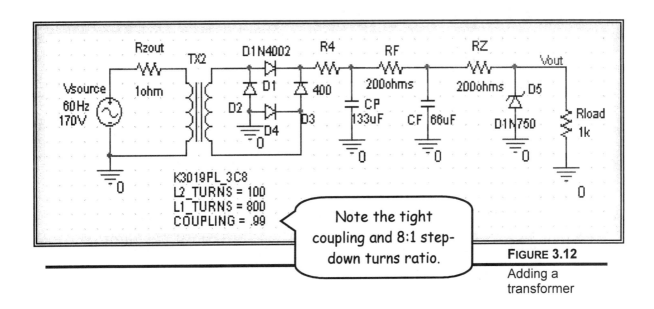

FIGURE 3.12

Adding a transformer

22. For each of the following malfunctions to the transformer circuit of Figure 3.12, predict the approximate output voltage waveform. Make the change to the circuit and compare to the PSpice-generated result.

    a. Diode *DR1* opens

    b. *CP* opens

    c. *RZ* shorts

Exercises

1. Perform a heat analysis on various components (Rs and diodes) in the circuit. Compare to specification sheet and rated values. (*Hint*: Graph $V*I$ or I^2R.)

2. Perform a Fourier analysis on the waveforms at various stages from input to output. (See Volume I, Chapter 19.)

3. Design a regulated 9.4V power supply. (*Hint*: Place two D1N750 zener diodes in series.)

Questions and Problems

1. Besides the power supply and load, what two components are required for a peak rectifier?

2. Fill in the blanks below with *resistor* or *capacitor*.

   With an RC low-pass filter, the DC component appears primarily across the _____ and the AC component (ripple) appears primarily across the _____.

3. Referring to Figure 3.8, circle all the following processes that will decrease the ripple:

    a. increase CP
    b. decrease CP
    c. increase CF
    d. decrease CF

4. Besides regulating the voltage, why does the zener voltage regulator circuit also further reduce the ripple? (*Hint*: How does the zener *RZ* combination act as a voltage divider?)

5. Why is an *LC* filter more efficient than an *RC* filter?

6. Why is the surge current greatest during the first cycle?

4

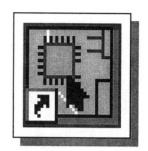

Clippers,
Clampers,
and Multipliers

Component Initialization

Objectives

- *To design and analyze a variety of clippers, clampers, and multipliers*
- *To initialize components*

Discussion

Three of the most common applications of the diode are *clippers*, *clampers*, and *multipliers*.   They are defined as follows:

- A *clipper* is a combination of diodes and resistors that limits the magnitude of a time-domain waveform.

- A *clamper* is a series combination of a diode and capacitor that adds a DC component to a time-domain waveform.

- A *multiplier* is a combination of diodes and capacitors yielding a DC voltage that is a multiple of the peak input voltage.

Component Initialization

Quite often we are interested in a circuit's steady-state response. However, based on the results of the power supply of Chapter 3, a great deal of computing time is often needed just to reach steady state. One case is the multiplier circuit of this chapter. One solution is to *initialize* the capacitors to near full charge before computation begins.

Simulation Practice

Activity *CLIPPER*

Activity *CLIPPER* compares the input and output of simple and biased clipper circuits.

Hand Calculations

1.   Figure 4.1 shows a simple clipper and a *biased* clipper. In each case, predict the output time-domain waveforms for a +5V sine wave input. Sketch your predictions on the graphs of Figure 4.2.

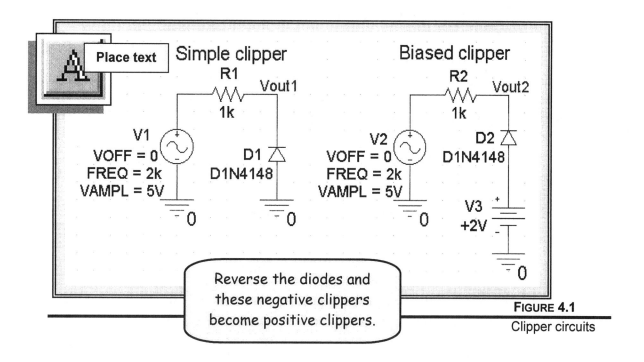

FIGURE 4.1

Clipper circuits

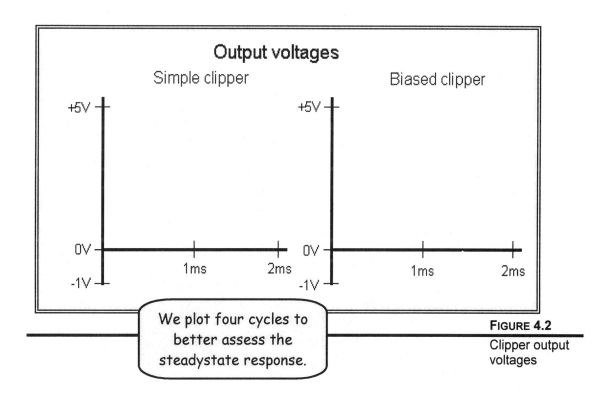

FIGURE 4.2

Clipper output voltages

PSpice Analysis

2. Create project *ccm* with schematic *CLIPPER* and page *PAGE1*.

3. Draw the circuits and set the simulation profile to *transient* from 0 to 2ms, with a step ceiling of 2μs.

4. Run PSpice and generate the curves of Figure 4.3. Did your predicted curves match the experimental (PSpice) curves?

Yes   No

Reminders

- To open a second window: **Window, New Window, Tile Vertically**.
- To open a second plot: **Plot, Add Plot to Window**.

For a complete description of new windows and plots, see *Simulation Note 13.2* (Volume I).

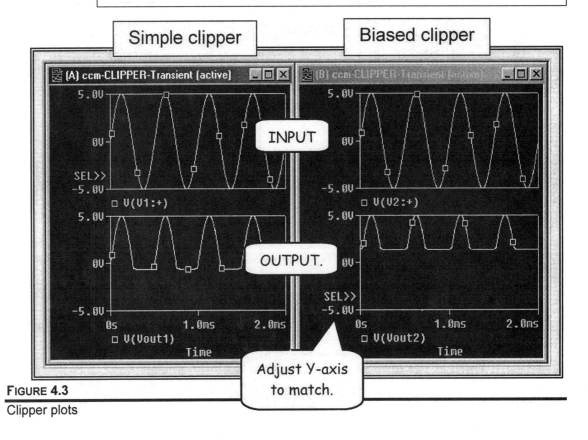

FIGURE 4.3

Clipper plots

Activity *CLAMPER*

Activity *CLAMPER* compares the input and output of simple and biased clamper circuits.

Hand Calculations

5. Figure 4.4 shows a simple clamper and a *biased* clamper. In each case, predict the output time-domain waveforms for a +5V sine wave input. Sketch your predictions on the graphs of Figure 4.5.

PSpice Analysis

6. Add schematic *CLAMPER* to project *ccm*, draw the circuits of Figure 4.4, and set the simulation profile to *transient* from 0 to 2ms, with a step ceiling of 2µs.

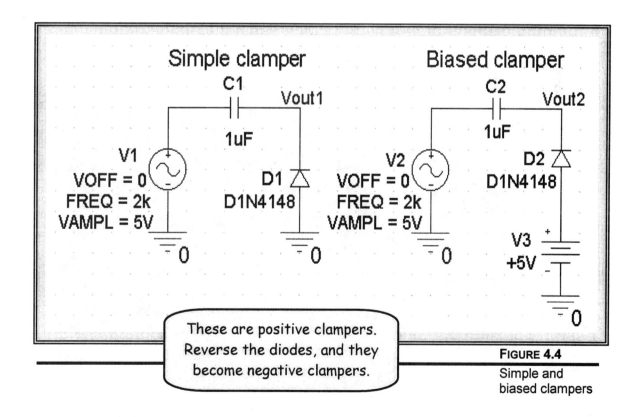

FIGURE 4.4

Simple and biased clampers

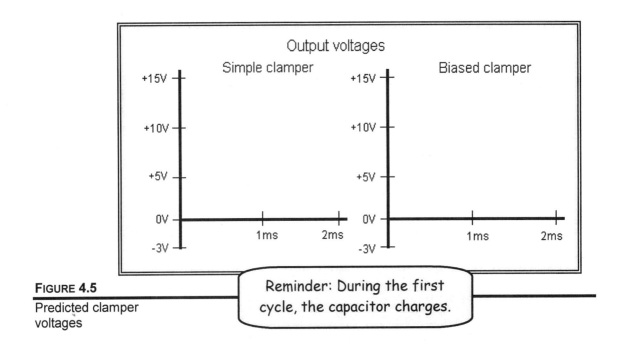

FIGURE 4.5

Predicted clamper
voltages

7. Run PSpice and generate the curves of Figure 4.6. Did your
predicted curves match the experimental (PSpice) curves?

Yes No

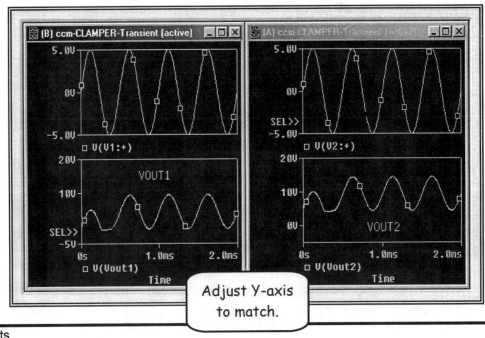

FIGURE 4.6

Clamper plots

Activity *MULTIPLIER*

Activity *MULTIPLIER* compares the input and output of a single-stage multiplier.

Hand Calculations

8. Figure 4.7 shows a common form of multiplier. In essence, it is a positive clamper followed by a peak rectifier. Predict the output voltage waveform and sketch on the graph of Figure 4.8.

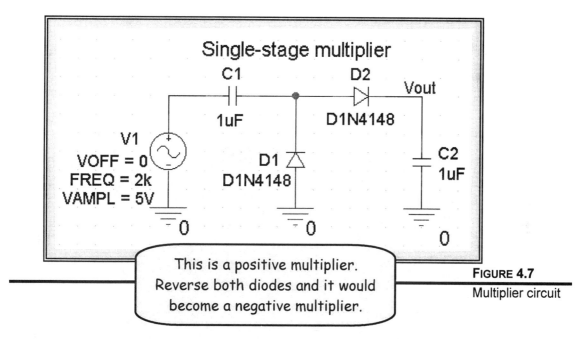

Single-stage multiplier

This is a positive multiplier. Reverse both diodes and it would become a negative multiplier.

FIGURE 4.7

Multiplier circuit

PSpice Analysis

9. Add schematic *MULTIPLIER* to project *ccm*, draw the circuit, and set the simulation profile to *transient* from 0 to 4ms, with a step ceiling of 4µs.

10. Run PSpice and generate the curves of Figure 4.9. Did your predicted curves match the experimental (PSpice) curves?

        Yes   No

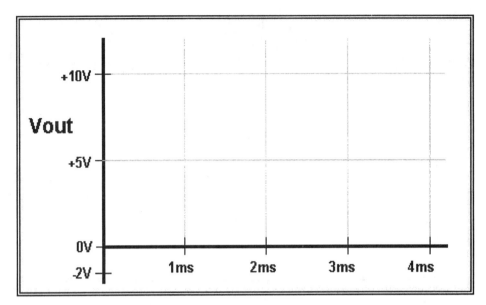

FIGURE 4.8

Predicted multiplier
waveform

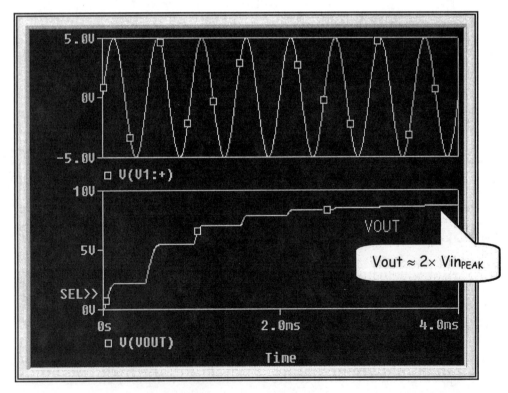

FIGURE 4.9

Multiplier response

PSpice for Windows

11. Chances are that step 10 revealed large differences between the predicted and actual (PSpice) waveforms, because the capacitors must be pumped up during the earlier cycles to approach their steady-state values.

Bring up the *Simulation Settings* dialog box and set up the *Run to Time* and *Start Saving Data After* fields as shown. This sets up the system to show the waveforms from approximately 50 to 55 cycles (25ms to 27.5ms) as it approaches steady state.

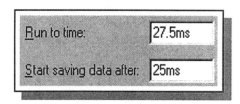

12. Run PSpice and generate the waveforms of Figure 4.10.

   • What is the average value of *Vout* between 25ms and 27.5ms? _____

   • Is *Vout* still increasing?

      Yes   No

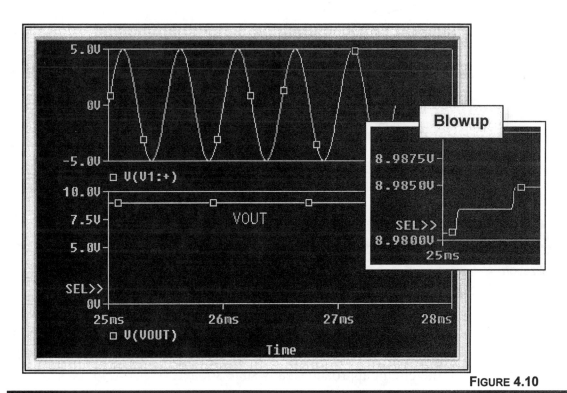

FIGURE 4.10

Multiplier response

Component Initialization

Another way to overcome the time it takes to reach steady state is to initialize the capacitors to nearly their final values at the beginning of calculations. This is most easily done with components *IC1* and *IC2*.

- *IC1* sets an absolute voltage with respect to ground.
- *IC2* sets a differential voltage.

13. Precharge capacitor *C1* to 4.4V and *C2* to 8.7V by placing parts *IC2* as shown in Figure 4.11. (We use 4.4V and 8.7V because of the barrier potentials of *D1* and *D2*.)

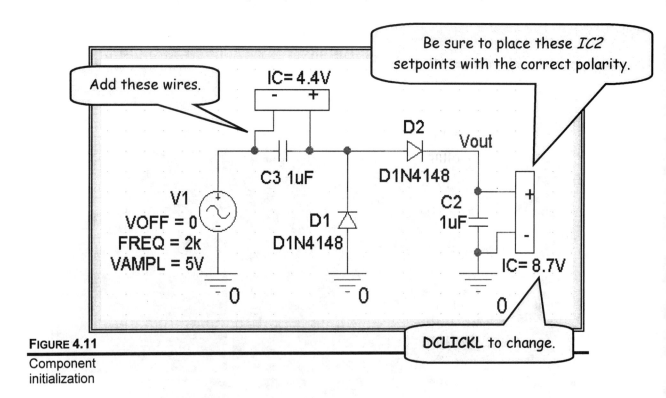

FIGURE 4.11

Component initialization

14. Change the transient display back to 0 to 4ms, display the new waveforms, and compare with the results of Figure 4.9. Does the circuit approach steady state sooner?

Yes   No

Advanced Activities

15. Using PSpice do a *surge current* analysis of the multiplier circuit of Figure 4.7. Do the numbers indicate that a surge-protection resistor is needed?

16. Test the clippers, clampers, and multiplier circuits of this chapter with square wave inputs.

Exercises

1. Design a digital-based clipper (limiter) circuit that limits an input waveform to the range from 0 to +5V.

2. Design the biased clamper that results in the output mystery waveform of Figure 4.12.

3. By using two back-to-back (mirror image) multipliers (doublers) of Figure 4.7, design a multiplier that increases the peak input voltage by a factor of 4.

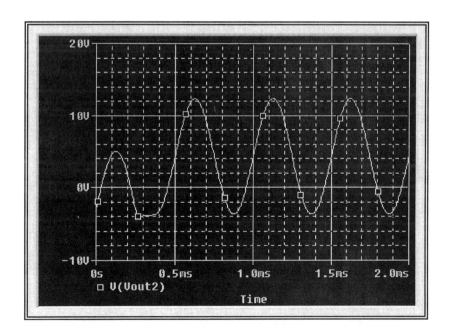

FIGURE 4.12

Mystery waveform

Questions and Problems

1. What two components are necessary for clamping, and how must they be arranged?

2. Show how to use a silicon diode (barrier = .7V) and a germanium diode (barrier = .3V) to create a clipper that limits an input waveform to the 0 to +1V range.

3. When voltage is multiplied, what happens to the current? Why?

4. Clamping is quite often unwanted. Referring to the circuit shown in Figure 4.13, how does resistor *R3* reduce the effects of clamping?

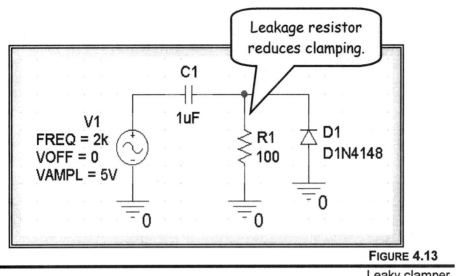

FIGURE 4.13

Leaky clamper

5.   Referring to Figure 4.7, why is capacitor *C2* pumped up over time? (Why does it take an infinite number of cycles for *C2* to reach full charge?)

5

The Analog Switch

Crash Studies

Objectives

- *To perform crash studies using a voltage-controlled switch*
- *To use a diode to simulate the effects of safety devices*
- *To unsynchronize an X-axis plot*
- *To set a watch alarm*

Discussion

This chapter's simulation study involves automotive safety. One of the most dangerous situations results from a quick stop, a crash. A crash is especially dangerous because of the great forces that can build up even at moderate speeds. To model a sudden stop, we use the circuit of Figure 5.1.

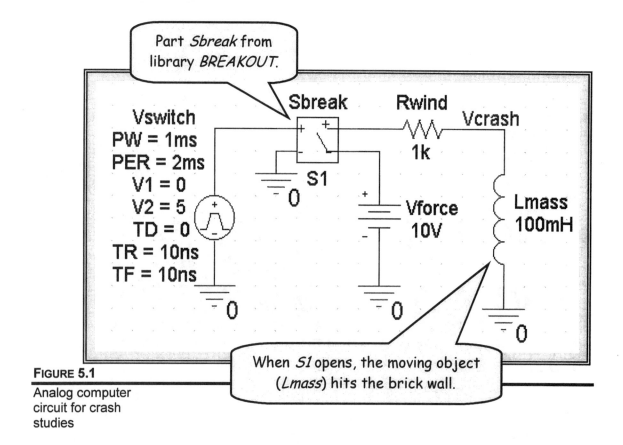

FIGURE 5.1

Analog computer
circuit for crash
studies

Component *S1* is a voltage-controlled switch. When the switch closes, *Vforce* is applied to *Lmass* and the car builds up velocity (current) until all the *Vforce* is used to overcome wind resistance (*Rwind* × I). When top speed is reached, maximum kinetic energy is contained in *Lmass* ($1/2LI^2$). When the switch opens (a crash occurs), the velocity is suddenly forced to zero and the back EMF ($V = L\Delta I/\Delta t$) simulates the very large crash forces.

To counter the large forces developed in a crash, we must dissipate the energy over a longer period of time with the use of safety devices, such as seat belts and air bags. To model such safety devices, we use a diode to separate the speed-up (accelerate) portion of the simulation from the slow-down (crash) portion.

Simulation Practice

Activity *SWITCH*

Activity *SWITCH* uses the analog computer of Figure 5.1 to simulate a crash by suddenly reducing the current to zero and noting the back EMF forces across inductor *Lmass*.

1. Create project *crash* with schematic *SWITCH*.

2. Draw the test circuit of Figure 5.1.   (Switch *S1* is part *Sbreak* from library *breakout.olb*.)

3. Set the simulation profile to transient from 0 to 4ms, with a step ceiling of 4μs.

4. Run PSpice and generate the curves of Figure 5.2.

    a. Does the car reach steady state velocity prior to the crash?

    > Yes   No

    b. Which component *stores* energy prior to the crash?

    > R   L

    c. How large are the crash forces (back EMF voltage) and when do they occur?

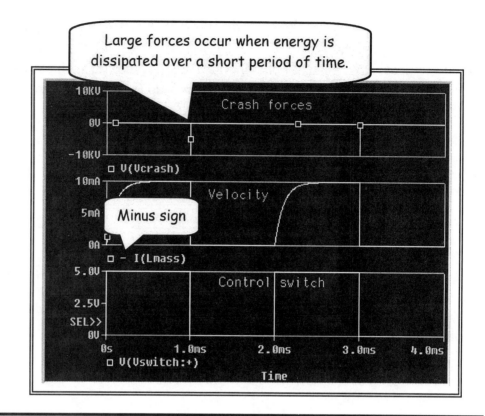

FIGURE 5.2

Crash study
results

5. Figure 5.2 shows us what happens when an object is stopped suddenly with no restraining devices used. To simulate the effects of an air bag, add the resistor/diode circuit of Figure 5.3. The back EMF current now has a pathway through *Rbag*, and the forces will be reduced because the energy will be dissipated over a longer period of time.

6. Test our new restraining system by generating the curves of Figure 5.4.

    a. Are the crash forces greatly reduced?

    Yes   No

    b. Are the crash force spikes slightly wider?

    Yes   No

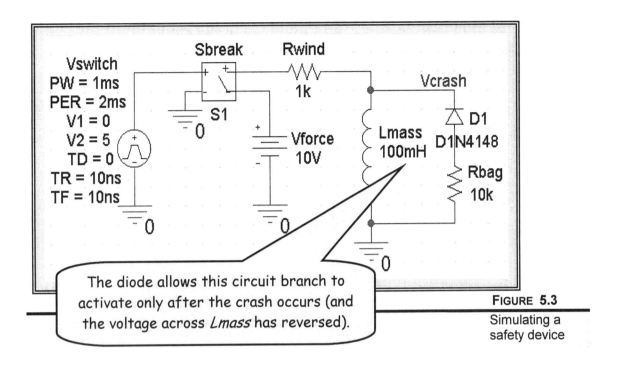

The diode allows this circuit branch to activate only after the crash occurs (and the voltage across *Lmass* has reversed).

FIGURE 5.3

Simulating a safety device

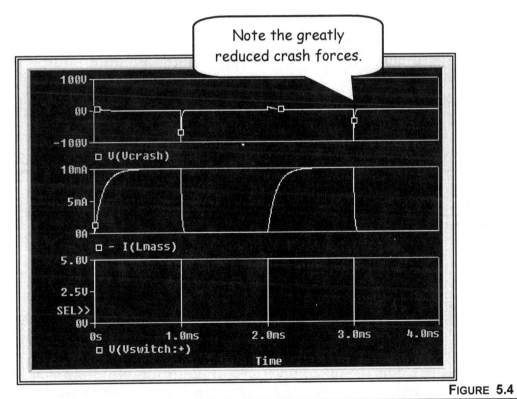

Note the greatly reduced crash forces.

FIGURE 5.4

Restraining system curves

PSpice for Windows

To better interpret the forces that occur during the moment of crash, we would like to zoom in on just the *crash force* plot, leaving the *control switch* and *velocity* plots as they stand.

One method is to uncouple (unsync) the crash force plot and give it an X-axis time range that is independent of the other plots.

7. Review *Simulation Note 5.1* to uncouple the crash force plot, then zoom in on the crash force spike and generate the waveform set of Figure 5.5.

   a. Have the forces been spread out over time?

      Yes   No

   b. Can we use the X-axis scroll bars to examine the entire original (unexpanded) waveform?

      Yes   No

   c. If the Y-axis is expanded, do scroll bars appear?

      Yes   No

Simulation Note 5.1
How do I uncouple plots?

When multiple plots are first generated, their X-axes are synchronized and any X-axis action affects them all.

To uncouple (unsynchronize) any of the top-most plots: **CLICKL** on plot to select (place SEL>>), **Plot**, **Unsynchronize X-axis**. (The bottom-most plot need not be unsynchronized as it is changed in the normal manner.)

We are then free to use any zoom, plot, or scroll technique on the uncoupled plot independent of all the others.

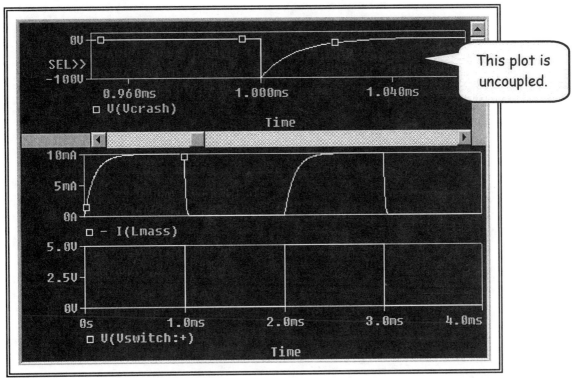

FIGURE 5.5

Unsynced crash
force plot

Watch Alarm

In many circuits, certain voltages may be critical. For example,
let's say that the maximum safe crash force corresponds to –80V.
During simulation, we wish to know if and when this value is
exceeded.   That is the job of a *watch alarm*.

8.   Following the steps of *Simulation Note 5.2*, first set a watch
point, as shown in Figure 5.6. Next, set the alarm trigger value
to –80V, run PSpice, and note the result in the *Watch* window of
Figure 5.7.

In the space below, record the watch alarm time and the
watch value (*V(Vcrash)*).

Watch alarm time = _____

Watch value (@ alarm time) = _____

Simulation Note 5.2
How do I set a watch alarm?

1. Place part *WATCH1* from library *SPECIAL* at the desired place on the schematic.

2. **DCLICKL** part *WATCH1* and set the *LO* and *HI* voltage values. (The system will halt if the voltage at the selected point reaches or exceeds the LO or HI values.)

3. Run PSpice. (Simulation will pause if and when the watch value is reached.)

4. Check the *Analysis* window to determine when the alarm occurred and the *Watch* window to determine the value reached.

5. **Simulation**, **Pause** (unpause) to allow the simulation to complete.

6. Plot curves as desired.

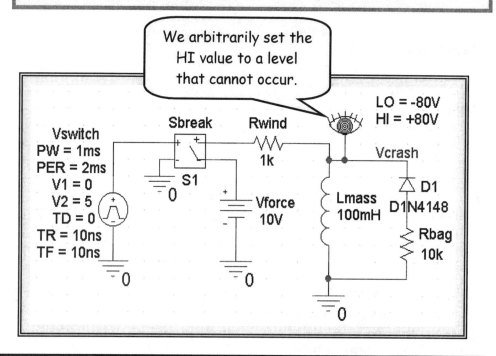

FIGURE 5.6

Setting a watch
alarm

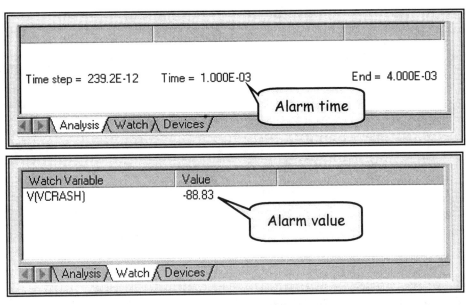

Time step = 239.2E-12   Time = 1.000E-03   End = 4.000E-03

Alarm time

Analysis / Watch / Devices /

Watch Variable	Value
V(VCRASH)	-88.83

Alarm value

Analysis / Watch / Devices /

FIGURE 5.7

Alarm value
reached

9. Allow the simulation to complete (**Simulation, Pause**) and
    display *V(Vcrash)*.   Did the forces first exceed −80V at the
    time indicated in step 8?

    Yes   No

10. Plot curve *Vcrash* and verify the watch values of Figure 5.7.

11. Because the watch alarm indicated a dangerous situation at
    1ms, lower *Rbag* to 7k (to dissipate the stored kinetic energy
    faster) and repeat the simulation.   Is the new design safe now?

    a. Did the watch alarm stay silent?

       Yes   No

    b. Are the crash forces always less than −80V?

       Yes   No

Advanced Activities

12. Using the circuit of Figure 5.3, create the demonstration graph of Figure 5.8.

> Figure 5.8 demonstrates all of the multiple-curve display methods available under *Probe*: *multiple windows*, *multiple plots*, *multiple Y-axes*, *unsynched plots*.
>
> Be aware that a darkened border refers to a selected window, SEL>> refers to a selected plot, and >> refers to a selected Y-axis. **CLICKL** on window, plot, or Y-axis, respectively, to select.

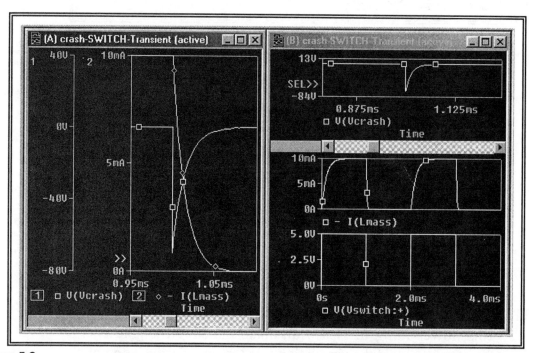

FIGURE 5.8

Multiple-plot
demonstration

13. Based on the data presented by our demonstration graph, does the relationship between velocity and force obey Newton's second law ($F = m\Delta V/\Delta t$)? (*Hint*: Does $V = LdI/dt$?)

Yes   No

14. Repeat the crash studies using the open/close switches of Figure 5.9. (Switches U1 and U2 are parts *Sw_tClose* and *Sw_tOpen* from library *eval.slb*.)

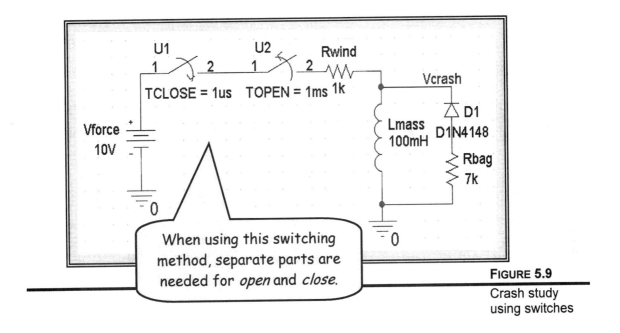

When using this switching method, separate parts are needed for *open* and *close*.

FIGURE 5.9

Crash study using switches

Exercises

1. By integrating the velocity to give distance (S operator), determine how far an object travels during the crash period. (How does this affect the design of the air bag?)

2. Perform a power/energy study of both crashes (with and without *Rbag*). How much energy was absorbed by *Rbag*? Suggestion: Use the "s" (integrate) operator to determine the area under a power curve, which equals energy.

3. In step 11 we found that lowering the value of *Rbag* reduced the maximum crash forces. Perform a study of this effect, and determine why there is a limit on how low we can take *Rbag*. (*Hint*: Lowering *Rbag* increases travel distance.)

To help your study, refer to the following equation, which gives the distance traveled by an applied force. Based on this equation, you may wish to plot graphs of $s(s(V(Vcrash)/100mH)))$ or $s(I(Lmass))$.

$$\text{Distance} = \int Velocity\ dt = \int \int \frac{Force}{Mass} dt dt$$

Questions and Problems

1. Based on the results of this chapter, does switch $S1$ bounce? How could bouncing be simulated? (*Hint*: We have total control over the opening and closing of the switch.)

2. When *Rbag* is not present (Figure 5.1), where does the inductor (kinetic) energy go when a crash occurs?

3. To make the restraining system more effective (lower the crash forces still more), should *Rbag* (presently 10kΩ in Figure 5.6) be increased or decreased? Why? (*Hint*: See step 11.)

4. Referring to Figure 5.8, when the crash force is maximum, what is true about the current? (*Hint*: $V = LdI/dt$.)

5. Which of the following equations demonstrates Newton's second law of motion ($F = ma$, where $a = dVelocity/dt$)?

   a. $V = LdI/dt$
   b. $E = 1/2LI^2$
   c. $Q = CV$

6. What does our crash study tell you about the dangers of back EMF (the voltage caused by the presence of an inductor in a circuit that opens suddenly)?

Part 2

Bipolar Transistor Circuits

In Part 2 we move to bipolar transistors, one of the fundamental building blocks of electronics. The emphasis is on amplifiers and buffers, from small-signal to classes A, B, and C.

Of special note is the introduction of frequency-domain analysis.

6

The Bipolar Transistor

Collector Curves

Objectives

- *To display collector and base curves for a bipolar transistor*
- *To determine a transistor's Beta (β)*
- *To determine transistor temperature effects*

Discussion

The bipolar transistor of Figure 6.1 is a solid-state device with two PN junctions.   Physically it seems to be nothing more than two back-to-back diodes.

However, when the center P region was made thin and when it was *biased* as shown in Figure 6.1, it took on truly revolutionary properties: It became a *voltage-controlled current source* with a current gain (*Beta*) in the range of 10 to 1000. It was an invention that changed the world.

VCIS

In simple terms a bipolar transistor is useful in analog applications because it is a *voltage-controlled current source* (VCIS).   That is, the input master voltage (V_{BE}) controls an output slave current (I_C), regardless of the slave voltage (V_{CB} or V_{CE}).

Although these VCIS properties were not new (vacuum tubes have them), it was the very first time they appeared in an inexpensive solid-state device.

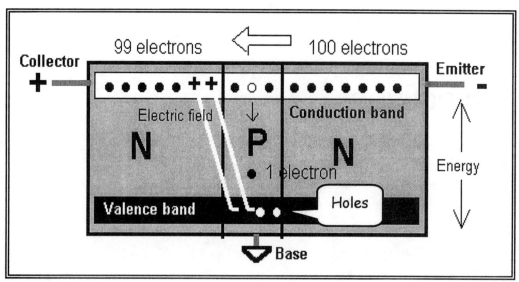

FIGURE 6.1

NPN transistor
physical configuration

Although it is not necessary to the circuit designer to know *how* a transistor achieves its VCIS characteristics, observe the following from Figure 6.1:

- The slave (collector-base) circuit is a reverse-biased PN junction that acts as a charged capacitor with an electric field between its plates. Therefore, no output current (I_C) flows as a result of output voltage (V_{CB}). Since the output current is independent of the output voltage, the collector is a current source (*IS*).

- The master (base-emitter) circuit is a forward-biased PN junction in which the input voltage (V_{BE}) pushes electrons into the P region conduction band. Held up briefly by quantum mechanical action, most of the electrons diffuse between the plates of the capacitor, are swept up by the strong electric field, and become collector current (I_C). Only a small percentage fall into the base's valence band and become base current (I_B). The output current is voltage controlled (*VC*) by the input voltage.

When these two physical facts are put together (*VC* + *IS*), a transistor becomes a VCIS. (With an input voltage, you get an output current—the output voltage doesn't matter.)

VCIS Versus ICIS

Because the master circuit of a bipolar transistor is a forward-biased PN junction, the driving source must supply current to the transistor. Furthermore, the relationship between base current *in* and collector current *out* is quite linear.   For these reasons, a bipolar transistor is also called a *current-controlled current source* (ICIS).

Transistor Beta (β)

To measure the effectiveness of any control device, we determine *output* divided by *input*. Using current as our input/output variables, the transistor of Figure 6.1 has a current gain (*Beta*) of 99/1 (I_C/I_B). Because *Beta* (β) typically lies in the 10 to 1000 range, and because $I_E = I_B + I_C$, β is also approximately equal to I_E/I_B.

Saturation

Should the voltage between the collector and emitter (V_{CE}) drop below $\approx$ .3V, the resulting weak electric field between collector and base is no longer able to sweep 99% of the electrons into the collector. Instead the electrons fall into (and saturate) the base.

Therefore, the region from V_{CE} = 0 to approximately .3V is known as the *saturation* region. In the saturation region, the transistor is no longer a current source. (In the saturation region, I_C *does* depend on V_C.)

When used in an analog application, we generally stay away from the saturation region; when used in a digital application, we generally use the saturation region as one of the two logic states.

NPN Versus PNP

The test circuits of Figure 6.2 show the schematic symbols for both the NPN and PNP versions. The PNP transistor performs like the NPN, except that all voltages and currents are reversed. (Note the change in direction of the arrow between the base and emitter.) We will use these circuits to investigate the VCIS/ICIS characteristics of a bipolar transistor.

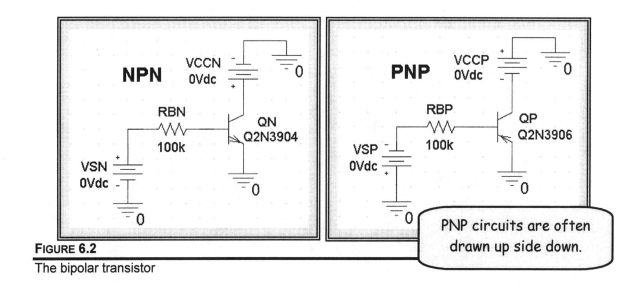

FIGURE 6.2

The bipolar transistor

Simulation Practice

Activity NPN

Activity NPN gives us a means of generating various bipolar transistor test plots, such as collector curves.

1. Create project *transistor* with schematic *NPN*.

2. Draw the NPN test circuit of Figure 6.2.   (The NPN transistor is part Q2N3904 from library *eval*.)

3. Set the simulation profile as follows:

    ▪ The *primary* sweep is a linear DC Sweep of VCCN from 0 to +10V in increments of .01V.

    ▪ The *secondary* (nested) sweep is a linear DC Sweep of *VSN* from 0V to +10V in increments of 2V.

4. Run PSpice and generate the collector curves of Figure 6.3.

    a. In the current source region, are the curves reasonably flat? Does this prove that the collector is a current source (*IS*)?

    >   Yes   No

    b. To move from one curve to the next, we change the value of V_{BE} (or I_B).   Does this prove that the collector current is voltage controlled (or current controlled)?

    >   Yes   No

    c. Does the transistor saturate when the collector voltage (*VCCN*) drops below approximately .3V?

    >   Yes   No

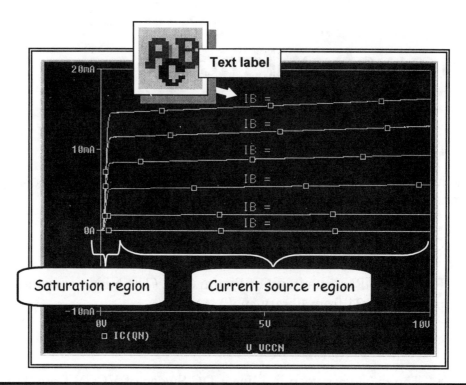

FIGURE 6.3

Collector curves

5. Add another Y-axis (or plot) to your graph, and display a set of six base current (*IB(QN)*) curves. Realizing that the new curves correspond one-to-one to the original curves, fill in the corresponding "IB =" values for each on Figure 6.3. (Use the margin and arrows, if necessary.)

6. Using the displayed data in the current source region, calculate a typical value of β (current gain) for the top curve.

$$\beta \text{ (top curve)} = I_C/I_B = \underline{\hspace{3cm}}$$

Collector Impedance

By definition, the AC impedance (Z_{AC}) of an ideal current source is infinity. Based on the equation below, this means that a finite change in voltage results in no change in current. Because a real-world transistor is not ideal, its Z_{AC} will be considerably less than infinity.

$$Z_{AC} = \Delta V_{CCN}/\Delta I_C$$

7. Labeling the curves 1 through 6, starting from the bottom, determine the AC collector impedance ($Z_{AC} = \Delta V/\Delta I$) in the current-source region for curves 2 and 6. (*Hint*: Figure 6.4 shows how to determine collector impedance for curve four.)

Z_{AC} (curve 2) = _____ Ω

Z_{AC} (curve 6) = _____ Ω

As we move from bottom to top, does the collector impedance decrease?

Yes   No

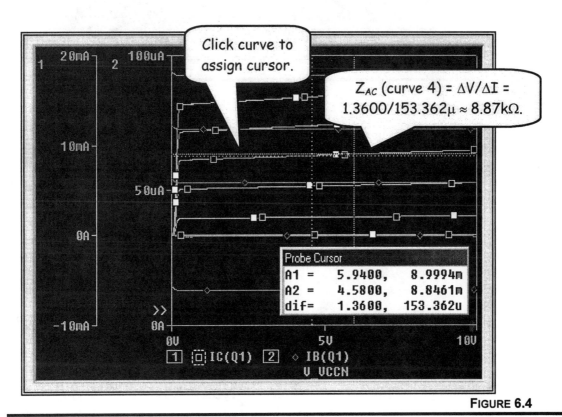

FIGURE 6.4

Determining collector impedance

Saturation

8. Expand the saturation region of the impedance graph of Figure 6.4 and generate the plot of Figure 6.5.

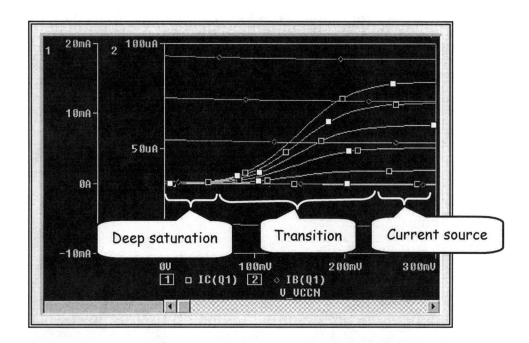

9. Based on the results of Figure 6.5, determine the value of *Beta* for the top-most curve at V_{CCN} = 300mV, 150mV, and 50mV. (Hint: I_B remains constant.)

	300mV	**150mV**	**50mV**
Beta			

Does β decrease (more electrons fall into the base) as we go from the current source region to deeper and deeper into the saturation region?

Yes   No

Base (Master) Curves

10. The collector curves of Figure 6.3 do not *directly* show the VCIS transconductance relationship. (How does the input *voltage* affect the output *current*?) To see this relationship, generate the master curve of Figure 6.6. (If necessary, refer to the Process Summary that follows.)

Process Summary for Master Bipolar Curve

- The DC Primary sweep variable is *VSN*, from 0 to +10V in increments of .01V. (The DC Secondary sweep is disabled and *VCCN* is set to +10V.)

- The X-axis variable is V_{BE} (*V(QN:B)*), and the Y-axis variable is I_C (*IC(QN)*).

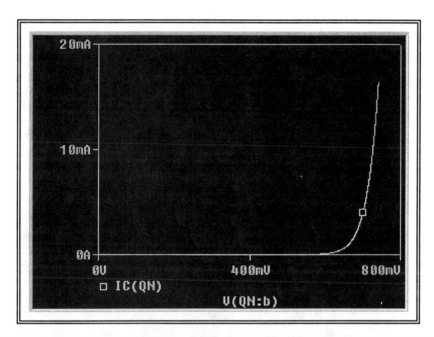

FIGURE 6.6

Collector current versus base voltage

11. As expected, Figure 6.6 shows the characteristics of a forward-biased PN junction.

    a. Is it similar to the diode curve of Chapter 1?

        Yes   No

    b. Is it only valid when the transistor is biased in the current source region?

        Yes   No

Temperature Effects

> Solid-state active devices are notoriously sensitive to temperature. Of all the temperature-related variables of a bipolar transistor, *Beta* is one of the most susceptible.

12. Recreate the graph of Figure 6.7, which shows how β changes with temperature. (If necessary, refer to the Process Summary below.)

> ### Process Summary for Temperature Effects of β
>
> The primary sweep variable is temperature, using a linear DC Sweep from –50°C to +50°C in increments of 1. (The secondary sweep is disabled; VCCN and VSN are both set to +10V.)

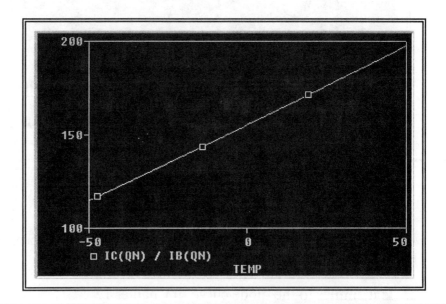

FIGURE 6.7

Beta as a function of temperature

13. By how much does β rise for each 1°C increase in temperature?

$$\Delta\beta/\Delta T = \text{slope} = \underline{\hspace{3cm}}$$

Advanced Activities

14. Generate a set of collector curves for the PNP circuit of Figure 6.2.   Based on your results, what is the difference in operation between an NPN and a PNP transistor?

15. Pick several points on the collector curves of Figure 6.3 and determine the amount of power dissipated by the transistor.

Exercises

1. Design a current source of 8mA using each of the following:

   - a 3904 NPN transistor
   - a 3906 PNP transistor

2. Compare the characteristics of a Q2N2222 NPN transistor with the Q2N3904 of this chapter.

Questions and Problems

1.   What is a *current source*?

2.   If for every 100 electrons that are pulled into the emitter 98 go on to the collector, what is the *Beta*?

3. With the emitter grounded, what is the approximate minimum collector voltage that will result in normal (nonsaturated) operation?

4. For the circuits of Figure 6.2, place "base-emitter" or "collector" in the correct spaces below.

   VCIS means that the input _____ voltage controls the output _____ current, regardless of _____ voltage.

5. A transistor's *alpha* is equal to I_C/I_E. If *Beta* for a given transistor is 200, what is *alpha*?

6. Based on the results of procedure step 7, what is a typical (average) value for Z_C (AC collector impedance in the current-source region)? Is there a difference between DC impedance and AC impedance?

7. When the temperature rises, Beta

   a. goes up.
   b. goes down.

8. When using the transistor schematic symbol, the arrow points in the direction of emitter

   a. electron flow.
   b. conventional flow.

9. When a transistor enters saturation, V_C is approximately .3V higher than the emitter. However, V_B is .7V higher than the emitter. Does this mean that the collector can actually be lower in voltage than the base (by as much as .4V) and still receive 99% of the electrons entered from the emitter? (*Hint*: Reformat Figure 6.3 so the X-axis is V_{CB}, rather than V_{CE}.)

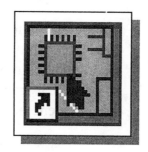

Bipolar
Biasing

Temperature Stability

Objectives

- *To analyze several bipolar transistor biasing circuits.*
- *To compare temperature stability of biasing circuits.*
- *To perform a DC sensitivity analysis on bias circuits.*

Discussion

In analog applications, a bipolar transistor is primarily used as an amplifier or buffer. Because analog applications usually involve an AC signal, we must *bias* the transistor. (As we will see in Chapter 16, a bipolar transistor used in a digital application is a *switch* and does not generally require biasing.)

To bias a bipolar transistor is to use a DC voltage to place its *quiescent* (Q) *point* at an appropriate place in the master curve. When properly biased, the superimposed AC signal will have room to operate on both its positive and negative cycles. A typical Q point for the 3904 NPN bipolar transistor is shown in Figure 7.1.

Because of a transistor's sensitivity to voltage (beyond the .7V knee), we usually establish the Q point by designing for the desired *current*. For example, as shown in Figure 7.1, if we design for a quiescent collector current of 10mA, the corresponding base-emitter voltage is automatically set.

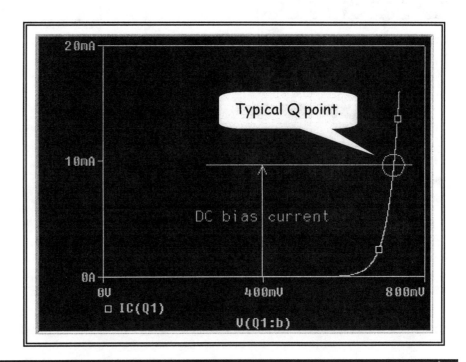

FIGURE 7.1

NPN transistor biasing
and Q points

There are many trade-offs involved in the design of a bias circuit. Several of the most important are *cost*, *temperature stability*, *sensitivity to tolerances*, and *number of power supplies required*. It follows that there is no single bias circuit that is best for all, but there are several that are preferred.

Stability

Stability is a temperature problem. Will a circuit that operates properly at room temperature continue to operate properly when the temperature rises (and the value of *Beta* changes, for example)?

To perform a stability analysis, we sweep through a range of temperatures and note the change in Qpoint values.

DC Sensitivity

Sensitivity is a tolerance problem. For example, we might use DC sensitivity analysis to tell us how each resistor in the circuit affects the Q point voltage. If we find that the Q point is very sensitive to a particular resistor, we might want to reduce that resistor's tolerance from 10% to 1%.

To perform a DC sensitivity analysis, we first select an output variable (such as Q point voltage). By performing a linear analysis of all devices about the bias point, the sensitivity of the output variable to all the device values and model parameters will be calculated and sent to the *output file*.

Simulation Practice

Activity *BASE*

Activity *BASE* uses the test circuit of Figure 7.2 to study the characteristics of base biasing. The circuit is simple and inexpensive—but is it stable?

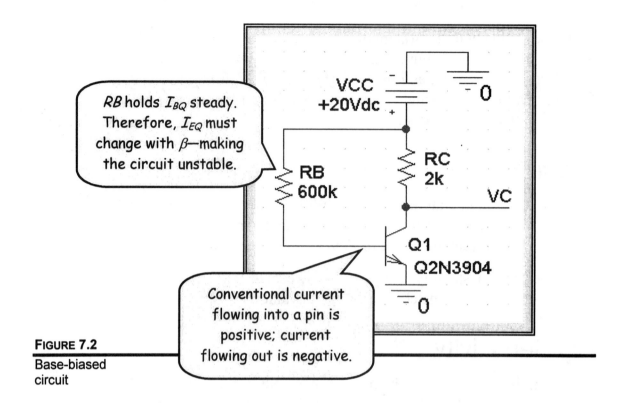

FIGURE 7.2

Base-biased
circuit

Hand Calculations

The following equations govern the bias point values for the base-biased circuit of Figure 7.2.

- $Beta = I_C / I_B =$ (approximately) I_E / I_B

- $V_{CC} = .7V + I_B R_B$

- $V_{CC} = V_C + I_C R_C$

1. Assuming a *Beta* of 175, solve for the Q point and fill in the following values. (C = "collector" and CE = "difference between collector and emitter.")

$$I_{CQ} = \underline{\hspace{2cm}} \qquad\qquad V_{CEQ} = \underline{\hspace{2cm}}$$

PSpice Analysis

2. Create project *bias* with schematic *BASE*.

3. Draw the test circuit of Figure 7.2, and set the simulation profile to *Bias Point*.

4. Run PSpice and report the Qpoint current and voltage.   (If necessary, **PSpice**, **Bias Points**, **Enable Bias Voltage**, **Enable Bias Current**, or use the corresponding toolbar buttons.)

I_{CQ} = _____ V_{CEQ} = _____

5. Compare the theoretical values of step 1 with the experimental values of step 4.   Are they approximately the same?

Yes   No

Temperature Stability

Using just a single base resistor, the *base-biased* circuit is inexpensive—but is it stable?   A good measure of stability is to answer the question, What happens to the Q point current (I_{CQ}) as the temperature changes?

6. To determine Q point stability, generate the graph of Figure 7.3 (a linear DC Sweep of temperature from −50 to +50.)

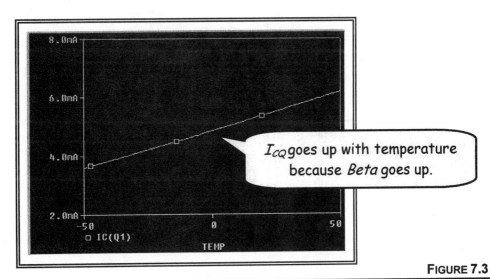

FIGURE 7.3

I_{CQ} as a function of temperature

7. Based on the results of Figure 7.3, it is clear that a base-biased circuit is not stable. As a quantitative measure of this instability, determine how Beta changes with each 1°C change in temperature.

$$\Delta I_{CQ} / \Delta T = \text{slope} = \underline{\hspace{3cm}}$$

Activity *VOLTAGE-DIVIDER*

Activity *VOLTAGE-DIVIDER* uses the test circuit of Figure 7.4 to demonstrate the most popular biasing circuit. At the modest cost of two additional resistors, it uses voltage-divider biasing to greatly improve stability.

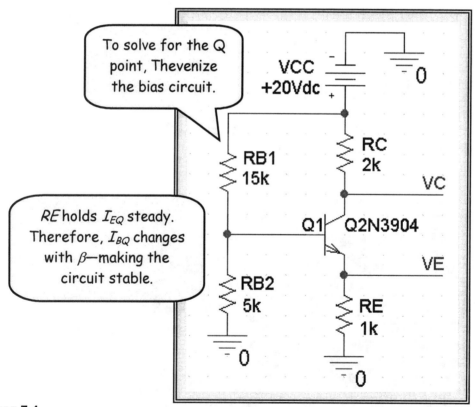

FIGURE 7.4

Voltage-divider bias

Hand Calculations

8. Determine the Qpoint values by hand calculation. (Hint: Thevenize the bias circuit.)

I_{CQ} = _____ V_{CEQ} = _____

PSpice Analysis

9. Add schematic *VOLTAGE-DIVIDER* to project *bias*, draw the circuit of Figure 7.4, and set the simulation profile to *Bias Point*.

10. Run PSpice, record each of the following, and compare to the hand-calculated values.

I_{CQ} = _____ V_{CEQ} = _____

11. By adding a DC Sweep simulation profile, generate a temperature curve similar to Figure 7.3, and record the following.

ΔI_{CQ} / ΔT = _____

Activity *OTHER*

Activity *OTHER* uses the test circuits of Figure 7.5 to test and compare two additional bias circuits: *Collector-feedback* biasing (uses negative feedback to improve stability) and *Emitter* biasing (uses a split power supply).

Hand Calculations

12. Calculate bias point values for the two circuits of Figure 7.5.

Collector feedback I_{CQ} = _____ V_{CEQ} = _____

Emitter biasing I_{CQ} = _____ V_{CEQ} = _____

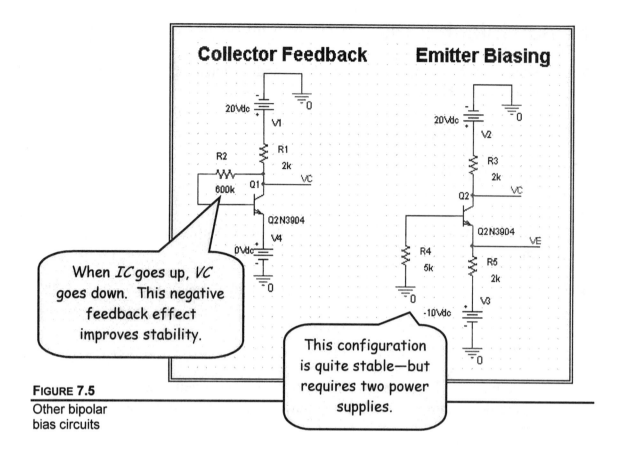

FIGURE 7.5

Other bipolar
bias circuits

PSpice Analysis

13. Add schematic *OTHER* to project *bias*, draw the test circuits of Figure 7.5 (both on the same schematic is okay), and set the simulation profile to *Bias Point*.

14. Use PSpice to determine each of the following.

Collector Feedback	Emitter Biasing
I_{CQ} = _____	I_{CQ} = _____
V_{CEQ} = _____	V_{CEQ} = _____
$\Delta I_{CQ}/\Delta T$ = _____	$\Delta I_{CQ}/\Delta T$ = _____

15. To better compare the temperature stability properties of the four bias circuits, plot all of them on a single graph (as shown in Figure 7.6).

> To combine data files: From Probe, **File**, **Append Waveform**, **DCLICKL** on desired data file, **Do not skip sections**. Repeat for each data file you wish to append.

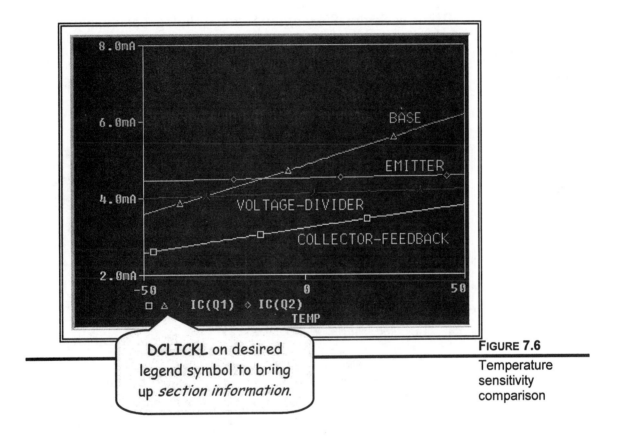

FIGURE 7.6

Temperature sensitivity comparison

16. Based on the results of Figure 7.6, what conclusions can you draw about the relative temperature stability of the four types of bias circuits?

DC Component Sensitivity

> Reminder: *Stability* is a temperature problem; *sensitivity* is a tolerance problem.

17. Bring back the voltage-divider bias circuit of Figure 7.4. Be sure to make *VOLTAGE-DIVIDER* the root schematic and *VOLTAGE-DIVIDER-Bias point* the active simulation profile.

18. Bring back the *Simulation Settings* dialog box (analysis type *Bias Point*). As shown by Figure 7.7, enable the *Perform Sensitivity Analysis* option and fill in the *Output variable(s)* field with all the variables on which a sensitivity analysis is to be performed [such as *V(Vc,Ve)*], **OK**.

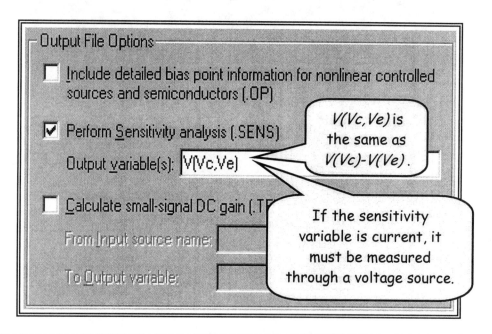

FIGURE 7.7

Sensitivity Analysis
dialog box

19. Run PSpice and bring up the *DC Sensitivity Analysis* portion of the output file (Table 7.1). The top of the table indicates that all values refer to the Q point voltage ($V(V_C, V_E)$), the voltage difference between collector and emitter.

```
     DC SENSITIVITY ANALYSIS      TEMPERATURE =     27.000 DEG C
           DC SENSITIVITIES OF OUTPUT V(VC,VE)
    ELEMENT          ELEMENT          ELEMENT        NORMALIZED
     NAME            VALUE          SENSITIVITY      SENSITIVITY
                                   (VOLTS/UNIT)    (VOLTS/PERCENT)
     R_RC          2.000E+03        -4.160E-03       -8.320E-02
     R_RB1         1.500E+04         7.287E-04        1.093E-01
     R_RB2         5.000E+03        -2.132E-03       -1.066E-01
     R_RE          1.000E+03         7.994E-03        7.994E-02
     V_VCC         2.000E+01         2.712E-01        5.424E-02
                                Q_Q1
     RB            1.000E+01         7.246E-05        7.246E-06
     RC            1.000E+00         1.936E-05        1.936E-07
     RE            0.000E+00         0.000E+00        0.000E+00
     BF            4.164E+02        -2.989E-04       -1.244E-03
     ISE           6.734E-15         2.904E+13        1.956E-03
     BR            7.371E-01         1.597E-10        1.177E-12
     ISC           0.000E+00         0.000E+00        0.000E+00
     IS            6.734E-15        -3.421E+13       -2.304E-03
     NE            1.259E+00        -3.346E+00       -4.213E-02
     NC            2.000E+00         0.000E+00        0.000E+00
     IKF           6.678E-02        -2.883E-01       -1.925E-04
     IKR           0.000E+00         0.000E+00        0.000E+00
     VAF           7.403E+01         4.211E-04        3.118E-04
     VAR           0.000E+00         0.000E+00        0.000E+00
```

PSpice, View Output File
to bring up the output file.

TABLE 7.1

DC sensitivity
data.

20. Next, we observe that the data (Table 7.1) is divided into two sections: resistor and voltage-source components at the top, and transistor parameters at the bottom.

a. How many R and V components affect $V(Vc,Ve)$? _____

b. How many transistor parameters affect $V(Vc,Ve)$? _____

21. To interpret the table, look at the first line. It tells us that, for every 1Ω change in the $2k\Omega$ value of R_C, $V(Vc,Ve)$ will change by $-.00416V$. Or, for every 1% change in R_C (20Ω), $V(Vc,Ve)$ will change by $-.0832V$.

As an example, if R_C changes from $2k\Omega$ to $2.2k\Omega$ (10%), enter the expected change in $V(Vc,Ve)$ below (Reminder: $V(Vc,Ve) = V_{CEQ}$):

ΔV_{CEQ} (from table) = _____

22. Using PSpice, record V_{CEQ} when $R\_RC = 2k\Omega$ and again when $R\_RC = 2.2k\Omega$. Enter the results below. Do they agree with those of step 21?

    ΔV_{CEQ} (from PSpice) = _____

23. By examining the data of Table 7.1, the output Q point voltage ($V(Vc,Ve)$) is most sensitive to a *percent* change in which resistor? Circle your answer.

    R_C R_{B1} R_{B2} R_E

    Would it make sense to reduce the tolerance of this resistor from the typical 10% to 1%?

    Yes   No

24. From Table 7.1, if the transistor's *maximum Beta* parameter (BF) changes by 10%, what change occurs in the output Q point voltage?

    $\Delta V_{CEQ} =$ _____

Advanced Activities

25. Generate an appropriate set of transistor collector curves and show and label each of the four Q-points.

26. Calculate and compare the power dissipated in the transistor for each of the four bias types.

Exercises

1. Design a bias circuit giving an $I_C = 10mA$ and having maximum temperature stability. (Hint: Which configuration is best?)

2. By changing the value of R_B, redesign the collector feedback circuit of Figure 7.5 to give *midpoint bias* operation ($V_C = 10V$).

3. Redesign any of the bias circuits of this experiment using a PNP (3906) transistor. (*Hint*: Following convention, draw the circuit upside down.)

Questions and Problems

1. What is the purpose of biasing a transistor?   (*Hint*: What would be the result if the transistor were not biased?)

2. Based on the bipolar examples of this experiment, which of the following results in the most stable circuit? (R_B refers to the base resistors and R_E refers to the emitter resistor.)

   (a) R_B small and R_E large

   (b) R_B large and R_E small

3. When biasing bipolar transistors, why is it easier to design for bias *current* (and let the bias voltage "tag along")?

4. What is the major disadvantage of emitter biasing (Figure 7.5)?

5. Describe the feedback process in the collector-feedback circuit of Figure 7.5. (Why is collector-feedback biasing more stable than base-biasing, although both use the same number of bias resistors?)

6. To improve the design of the voltage-divider bias circuit of Figure 7.4, you might lower the tolerance of two of the four resistors (for example, from 10% to 5%). Based on the results of the sensitivity analysis (Table 7.1), which two resistors would you select?

7. Why is voltage-divider biasing more stable than base-biasing?

CHAPTER

8

Bipolar Amplifier

Equivalent Circuits

Objectives

- *To analyze a common-emitter, small-signal amplifier*
- *To demonstrate the trade-off between linearity and gain*
- *To create DC and AC equivalent circuits*

Discussion

The small-signal amplifier circuit of Figure 8.1 starts with the very stable *voltage-divider* bias circuit of the last chapter. The signal is injected into the base at the front end, and the output is tapped off the collector at the back end. Because the emitter is AC-grounded, it is in the *common-emitter* configuration.

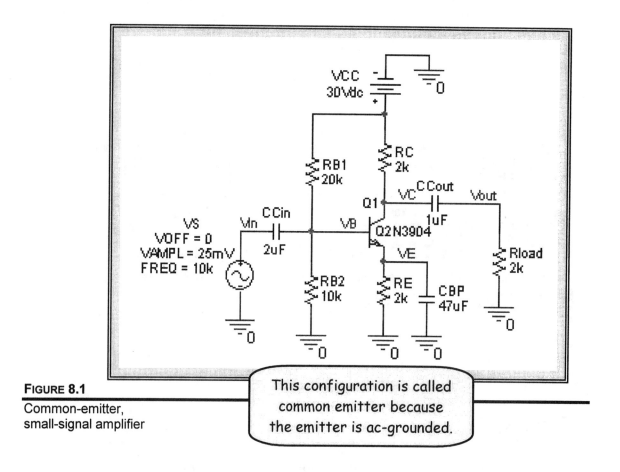

FIGURE 8.1

Common-emitter, small-signal amplifier

This configuration is called common emitter because the emitter is ac-grounded.

To better understand the action of the circuit, we apply the *superposition theorem* to divide its operation into two parts: a DC *equivalent circuit* (Figure 8.2), and an AC *equivalent circuit* (Figure 8.3). If we assume that all relationships are linear, then the total voltage or current at any circuit point is the algebraic sum of the DC and AC values. (DC variables are usually labeled with uppercase letters, and AC variables with lowercase letters.)

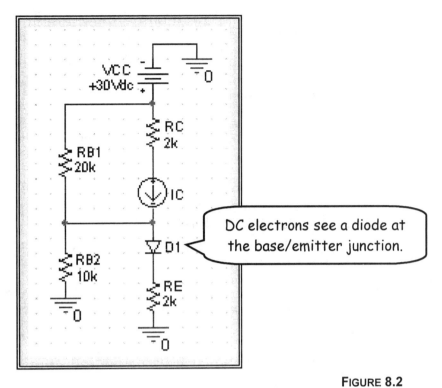

FIGURE 8.2

DC equivalent circuit

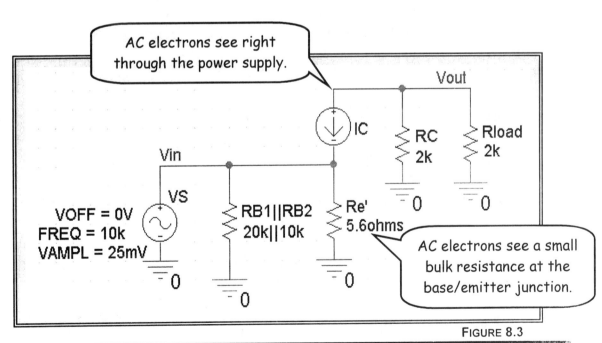

FIGURE 8.3

AC equivalent circuit

- **DC equivalent circuit**   The purpose of the DC equivalent circuit is to *bias* the transistor to its Q point. To extract the DC equivalent circuit, we open all capacitors, remove all AC sources, display the collector as a current source, and show the base-emitter (BE) junction as a diode.

The approximate DC values are as follows:

DC *Beta* = 175 (assumed value)

$I_{EQ} \cong I_{CQ} = (V_{TH} - .7V) / (R_E + (R_{B1}||R_{B2})/\beta)$
$(10V - .7V) / (2k\Omega + (10k\Omega||20k\Omega)/175) = 4.65mA$

$V_{CEQ} = V_{CC} - I_{CQ}(R_C + R_E) = 30V - 4.65mA(2k\Omega + 2k\Omega)$
        $= 11.4V$

- **AC equivalent circuit**   The purpose of the AC equivalent circuit is to amplify the signal.

To extract the AC equivalent circuit of Figure 8.3, we short all capacitors and DC power supplies, turn the collector into a current source, and show the BE junction as a bulk resistor.

The *approximate* AC values are as follows:

AC *Beta* = 175 (assumed to be the same as DC *Beta*)

Bulk resistance (re') $\approx$ 25mV / I_{EQ} = 25mV / 4.6mA = 5.4Ω

A (voltage gain) = $R_C||R_L$/re' = (2k || 2k) / 5.4Ω = 1k / 5.4Ω
        = 185

Z_{IN} (input impedance) = $R_{B1}||R_{B2}||(B \times re')$
= 10k||20k||(175 × 5.4Ω) = 900Ω

Z_{OUT} (output impedance) = $r_C||Z_{CURRENT\ SOURCE}$* = 2k||10k
        = 1.67k

* $Z_{CURRENT\ SOURCE}$ is from Chapter 6 and is assumed to be 10kΩ.

Coupling and Bypass Capacitors

The coupling capacitors (*CCin* and *CCout*) guide the AC in and out of the circuit without disturbing the DC biasing.   The bypass capacitor (*CBP*) increases the voltage gain by shorting the AC signal to ground (bypassing *RE*).   To perform their duties properly, the values of *CCin*, *CCout*, and *CBP* must be large enough to act as near shorts—but not too large because they can be expensive and bulky.

As an example, let's see how the value of *CCin* was determined. The first step is to Thevenize the amplifier and generate the equivalent circuit of Figure 8.4. We also assume (arbitrarily) that the lowest frequency of interest is 1kHz.

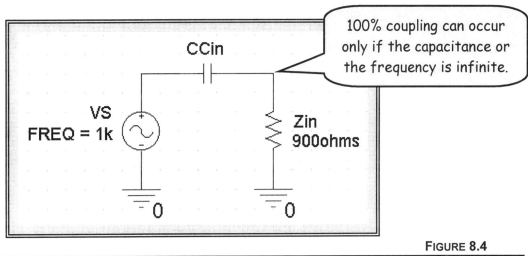

FIGURE 8.4

Equivalent circuit to isolate *CCin*

Because the capacitive reactance (X_{CCin}) can never be zero (unless f or C is infinite), we must choose a reasonable value for X_{CCin}. With a Z_{IN} of 900Ω, we arbitrarily choose a value of 90 (1/10 of Z_{IN}). The equation then becomes:

$$X_{CCin} = 1/2\pi fC = 1/(2 \times 3.14 \times 1k \times CCin) = 90\Omega$$

Solving for CCin yields approximately 2μF.

Values for the other two capacitors were determined in the same manner.

Experimentally Measuring Z$_{IN}$ and Z$_{OUT}$

Referring to the Thevenized version of the amplifier (Figure 8.5), we experimentally determine Z_{IN} by dividing V_{IN} by I_{IN}. We determine Z_{OUT} by measuring V_{OUT} with and without a load. Without a load (*RL* = infinity), $V_{OUT} = V_{TH}$; with a load, V_{OUT} is less than V_{TH}. We then use algebra and Kirchhoff's laws to determine Z_{OUT}.

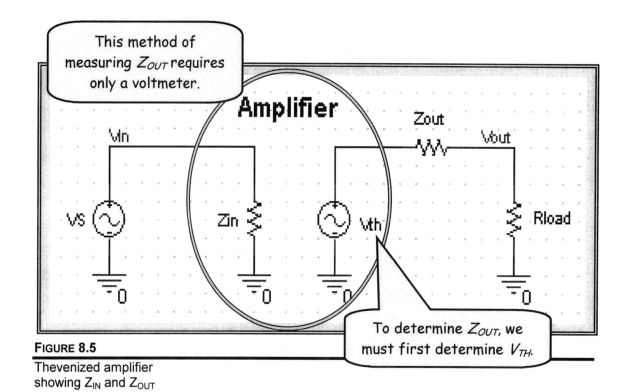

FIGURE 8.5

Thevenized amplifier
showing Z$_{IN}$ and Z$_{OUT}$

Time Domain vs Frequency Domain

In this chapter we use time-domain (transient) analysis to determine all amplifier characteristics, the type of analysis that might be done hands-on with an oscilloscope. In the next chapter, we will switch to frequency-mode analysis, the type of analysis that might be done hands-on with a spectrum analyzer. A complete analysis might include both types.

Simulation Practice

Activity *SMALLSIGNAL*

Activity SMALLSIGNAL analyzes the small-signal amplifier of Figure 8.1. We will calculate the "big three" (A, Z_{IN}, and Z_{OUT}), as well as several other important characteristics.

1. Create project *amplifier* with schematic *SMALLSIGNAL*, draw the circuit of Figure 8.1, and set the simulation profile to *Transient* from 0 to .2ms, with a step ceiling of .2µs.

DC Analysis

2. Use PSpice to perform a DC bias-point analysis and record each of the values listed below. (*Remember*: $V_{CEQ} = V_{CQ} - V_{EQ}$.)

    $I_{EQ} \cong I_{CQ} =$ _____ $V_{CEQ} =$ _____

    Are these values approximately the same (within 10%) as the calculations performed in the discussion?

        Yes   No

AC Analysis

3. Use PSpice to perform a transient analysis and generate the waveforms of Figure 8.6. Use the results to determine the following. (As stated in the discussion, determine Z_{OUT} by measuring V_{OUT} with and without a load.)

    A (V_{OUT} / V_{IN}) = _____

    AC *Beta* = I_C / I_B = _____

    $Z_{IN} = V_{IN} / I_{IN}$ = _____

    To average out the effects of nonlinear distortion, all measured values are peak-to-peak/2.

    Z_{OUT} = _____ (See discussion on Z_{IN} and Z_{OUT}.)

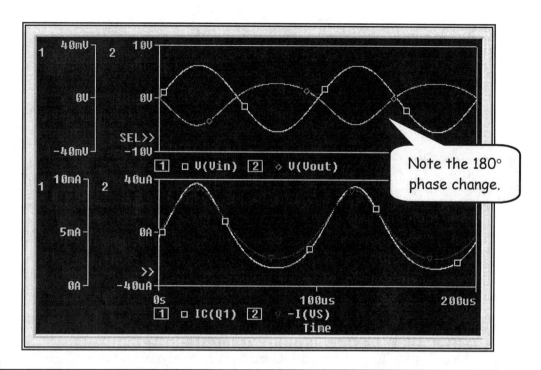

FIGURE 8.6

Small-signal amplifier
waveforms

4. Compare your experimental AC results from step 3 with the
    theoretical predictions made in the discussion. Are they
    generally the same (within 20%)? Comment on any significant
    differences.

5. In the discussion, we designed coupling capacitor *CCin* to drop
    less than 10% of the 25mV input signal at 1kHz. Using PSpice,
    verify our design calculations. (*Hint*: Be sure to lower the
    frequency to 1kHz, and be sure to record only the AC
    component.)

    Does *CCin* drop less than 10% of the input signal?

    Yes   No

Linearity

6. Look at the amplifier's output signal (Figure 8.6). Is it distorted (*nonlinear*)? Use the following equation to give a quantitative measure of the distortion.   (See Figure 8.7 for an example.)

$$\% \text{ distortion} = \frac{V_{PEAK}(difference)}{V_{PEAK}(average)} \times 100 = \underline{\hspace{2cm}}$$

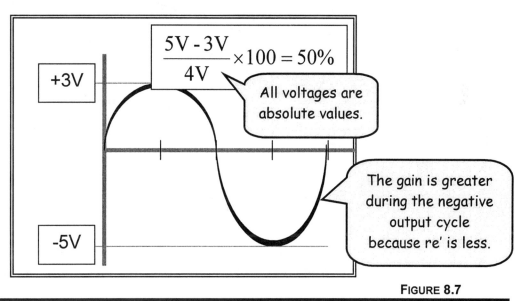

FIGURE 8.7

Percent distortion
calculation

7. To reduce distortion (at the expense of gain), we add the 100Ω *swamping* resistor (R_S), as shown in Figure 8.8. (*Note*: We increase V_S to 250mV to compensate for the reduced gain.)

8. Rerun PSpice. Swamping produced what percent change in each of the following?

Distortion reduced (%) = _____

Gain reduced (%) = _____

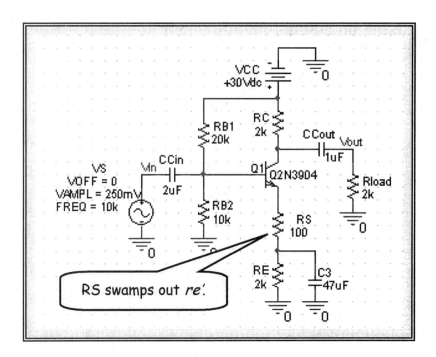

FIGURE 8.8

Swamped amplifier

Multiple Data Files

As we learned in the last chapter, the most direct way to compare waveforms from two separate circuits is to place both on the same window.

9. Display *Vout* for the swamped case.

10. Append *Vout* for the unswamped case to generate the graph of Figure 8.9. (**File**, **Append Waveform**, **DCLICKL** on the file name for the unswamped case, **Do Not Skip Sections**.)

Advanced Activities

11. Create the *common-base* (base is AC-grounded) circuit of Figure 8.10 by simply moving V_S from the base to the emitter. Measure *A* and Z_{IN}. Why did *A* remain the same but Z_{IN} dropped? (This configuration is often used in high-frequency applications because of its reduced input capacitance.)

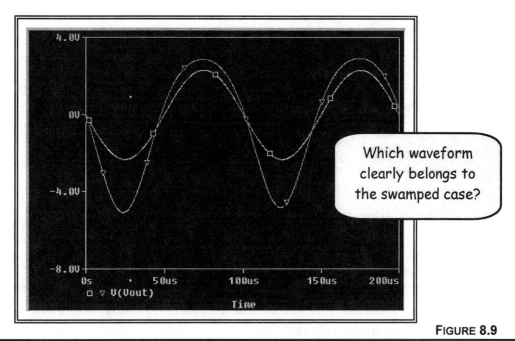

FIGURE 8.9

Combined files

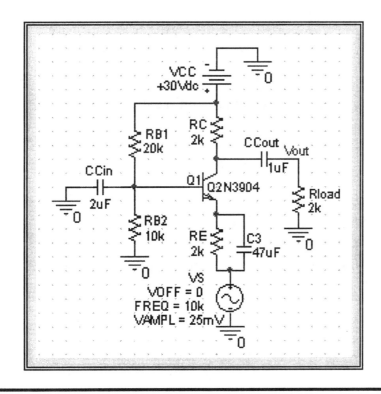

FIGURE 8.10

Common-base
configuration

PSpice for Windows

Exercises

1. Redesign the amplifier of Figure 8.1 using a PNP transistor. Are the major characteristics (A, Z_{IN}, Z_{OUT}, *linearity*) the same as those of the NPN transistor circuit?

2. Design a highly stable, highly linear *two-stage* common-emitter amplifier with an overall gain of 100.

Questions and Problems

1. To give an undistorted (linear) output signal, why is it important for the transistor to be a *current source*?

2. Why is the circuit of Figure 8.1 stable but nonlinear?

3. Regarding the amplifier circuit of Figure 8.1, what is the phase relationship between input and output?

4. Referring to Figure 8.8, what is the purpose or function of each of the following components?

    a. Capacitors *CCin* and *CCout*

    b. Capacitor *C3*

    c. Resistor *RE*

    d. Resistor *RS*

5. Referring to Figure 8.1, if the lowest frequency of interest were decreased from 1kHz to 100Hz, should we increase or decrease the values of the coupling and bypass capacitors?

6. What is the *power gain* of the circuit of Figure 8.8?   (*Hint*: Power gain = $A \times \beta$.)

7. What would happen to the voltage gain of the swamped circuit of Figure 8.8 if the bypass capacitor (*CBP*) opened?

8. Ignoring capacitive effects, what is the only difference between the *common-emitter* and *common-base* configurations? (*Hint*: How does Z_{IN} compare between the two?)

9

Bipolar Buffer

Frequency-Domain Analysis

Objectives

- *To analyze the common collector (emitter follower) buffer in the AC Sweep mode*
- *To perform parametric analysis on a model parameter*

Discussion

A *buffer* is first and foremost an *isolation* circuit. It follows that its most important characteristics are high Z_{IN} and low Z_{OUT}. Additional properties usually include a voltage gain near unity, moderate power gain, and a high degree of linearity.

One popular discrete circuit that achieves all the characteristics of a buffer is the *common-collector* (grounded collector) configuration of Figure 9.1.

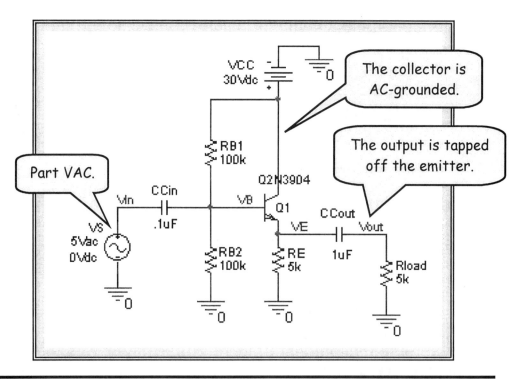

FIGURE 9.1

Common-collector
buffer

Note that the output is taken off the emitter rather than the collector. Because the output voltage is nearly equal to the input voltage, the circuit is also known as an *emitter follower* (the emitter voltage follows the base voltage). In short, the output is approximately equal to the input in both amplitude and phase.

Frequency-Domain Analysis

In the previous chapter we analyzed a small-signal amplifier in the time domain using transient analysis. In this chapter we analyze the buffer of Figure 9.1 by switching to the frequency domain and the *AC Sweep* mode.

A big advantage is that the properties of the circuit are displayed over a range of frequencies. This is vital for audio amplifiers because the frequency spectrum must match that of human speech and hearing.

THEORY

A complete and accurate frequency-domain theoretical analysis of the buffer of Figure 9.1 would involve elaborate equations of complex numbers. However, if we assume ideal *midband* characteristics, where all capacitors are short circuits, then the calculations are relatively simple.

The following equations determine the midband AC characteristics. (DC bias calculations are nearly the same as the previous chapter and are not included.)

Midband Calculations
$A = V_{OUT}/V_{IN} = (i_e \times R_E\|R_L)/(i_e \times (R_L\|R_E + re')) = 5k\|5k/(5k\|5k + 8.3)$ $= .997$ where $re' = 25mV / I_{EQ} = 25mV / 3mA = 8.3\Omega$
$Z_{IN} = R_{B1}\|R_{B2}\|\beta(re' + R_E\|R_L) = 100k\|100k\|175(8.3 + 5k\|5k) = 45k$
$Z_{OUT} = re'\|R_E = 8.3\|5k = 8.3\Omega$
$G(\text{power gain}) = (.5V_{OUT}^2/R_L)/(.5V_{IN}^2/Z_{IN}) = Z_{IN}/R_L = 45k/5k = 9$ $= 19.1dB$

As expected, voltage gain is nearly 1, input impedance is high, output impedance is low, and a modest amount of power gain is provided.

Looking at the gain equation (*A(buffer)*) we see that the linear term (5k‖5k) is much larger than the nonlinear term (8.3Ω), and therefore we predict a highly linear output waveform.

Simulation Practice

Unless you are fortunate enough to have a *spectrum analyzer*, the activities of this chapter are best carried out under PSpice alone.

Activity *COMMONCOLLECTOR*

Activity *COMMONCOLLECTOR* will verify the characteristics of the common collector buffer of Figure 9.1.

1. Create project *buffer* with schematic *COMMONCOLLECTOR*, and draw the circuit of Figure 9.1.

2. Set the simulation profile for a logarithmic (decade) *AC Sweep* from 10Hz to 100GHz at 100 points/decade.

Voltage Gain, Power Gain, and Bandwidth

3. Run PSpice and generate the voltage gain plot of Figure 9.2. (*T* = 1,000 *G*.)

Because all magnitude values are automatically peak, they can be plotted in combination form (such as *Vout/Vin*).

4. Upon viewing the results, give an approximate value for the midband voltage gain.

Midband voltage gain = _____

at frequency = _____

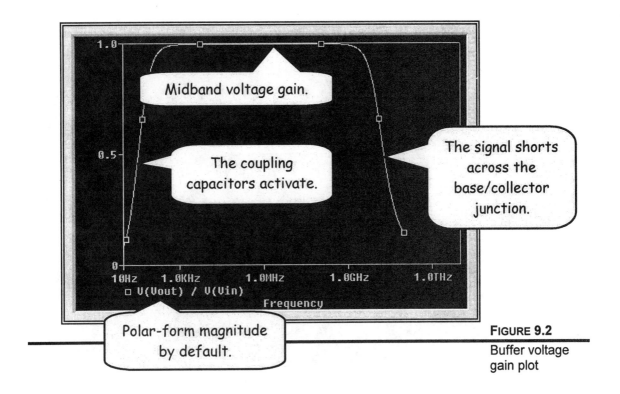

FIGURE 9.2

Buffer voltage gain plot

5. What is the approximate bandwidth of the buffer? (*Bandwidth* is the midband range of frequencies between the low- and high-frequency points where the gain drops to 3dB [70.7%] of its peak midband value.)

    Bandwidth = _____

6. Use the equation shown below to add a plot of *real* power gain (P_{OUT}/P_{IN}). The result is shown in Figure 9.3. (No *cos θ* is needed in the numerator because the voltage and current are in phase across a resistor.)

$$\text{Power gain (real)} = \frac{V(V_{OUT}) \times I(Rload)}{V(V_{IN}) \times I(VS) \times \cos(6.28/360 \times P(V(V_{IN})/I(VS))}$$

7. According to Figure 9.3, what is the approximate power bandwidth and midband power gain?

    Power bandwidth = _____

    Midband power gain = _____

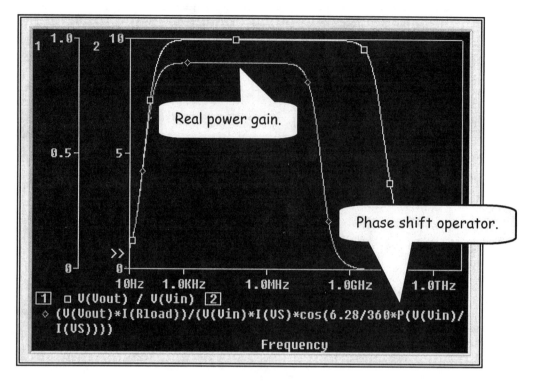

FIGURE 9.3

Adding power
gain

Phase Shift

8. Add a plot of phase change (between input and output) to your
    graph, as shown in Figure 9.4.   As expected (for an emitter
    follower), is the phase shift 0° at midband frequencies?

              Yes   No

Z_{IN} and Z_{OUT}

To complete the picture, we next display plots of the buffer's input
impedance (Z_{IN}) and output impedance (Z_{OUT}).

9. To show Z_{IN} as a function of frequency, display a plot of
    $V_{IN}/I(VS)$, as shown in Figure 9.5.   (Either add a new plot or
    delete the present curves.)

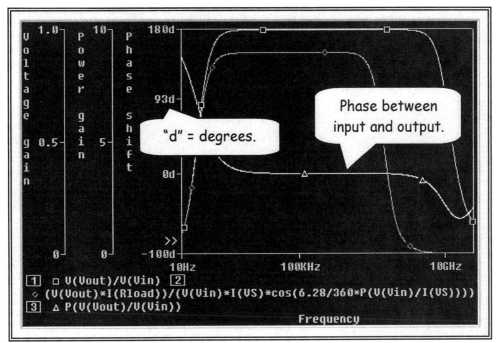

FIGURE 9.4

Adding phase

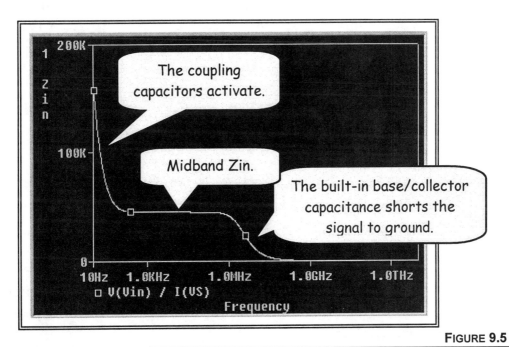

FIGURE 9.5

Plot of Z_{in}

10. What is the approximate midband Z_{IN}?

Z_{IN} (midband) = _____

11. To measure Z_{OUT} we move the source voltage from input to output, as shown in Figure 9.6.

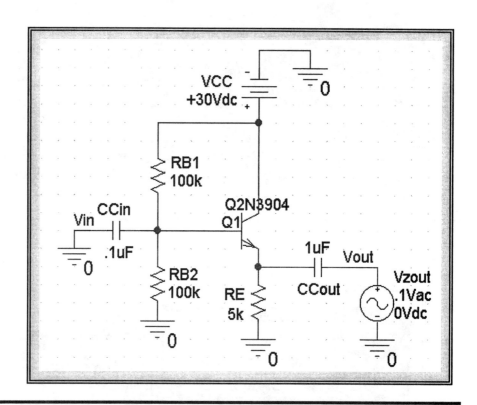

FIGURE 9.6

Determining Z_{OUT}

12. By plotting *V(Vout)/I(Vzout)*, generate the output impedance plot of Figure 9.7 and report the midband Z_{OUT}.

> The midband Z_{OUT} looks to be zero only because it is swamped by its high value at low frequencies. The cursor shows the actual value.

Z_{OUT} (midband) = _____

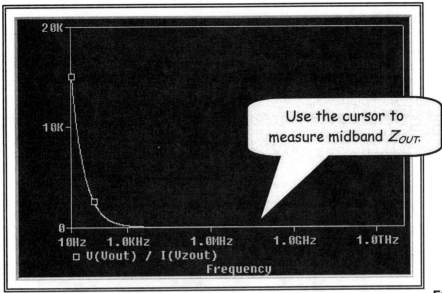

13. Does Z_{OUT} rise at low frequencies because *CCout* rises?

        Yes   No

14. To summarize our studies so far, complete the table below using values from the discussion and PSpice sections of the chapter. Are they approximately the same?

        Yes   No

	Calculated	PSpice
A		
G		
Z_{IN}		
Z_{OUT}		

Bode Plot

15. Return to the schematic of Figure 9.1, and use the *dB* operator to generate the Bode plot of Figure 9.8. Using the two "3dB down" points, measure the bandwidth. (How do the results compare with step 5?)

Bandwidth = _____

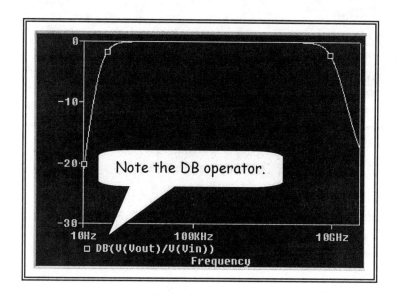

FIGURE 9.8

Voltage gain
in decibels

16. Based on all the previous results, does the circuit of Figure 9.1 have all the characteristics of a buffer?

Yes   No

17. To make sure that all the frequency-domain results are valid, change *VAC* to *VSIN* and run a transient analysis at a midband frequency (such as 100kHz). Is the output waveform free of clipping? (If not, all the previous results would be invalid, and the frequency-domain analysis would have to be rerun at a lower amplitude.)

Yes   No

Advanced Activities

Parametric Analysis of Model Parameters

Looking at the theoretical calculations in the discussion, Z_{IN} is highly dependent on *Beta*. To show this dependency, we will generate a family of Z_{IN} curves for a variety of *Beta* values. In the 3904 model, parameter *Bf* is the *ideal maximum forward Beta*.

18. For the buffer circuit of Figure 9.1, access the simulation settings dialog box and enter the values of Figure 9.9.

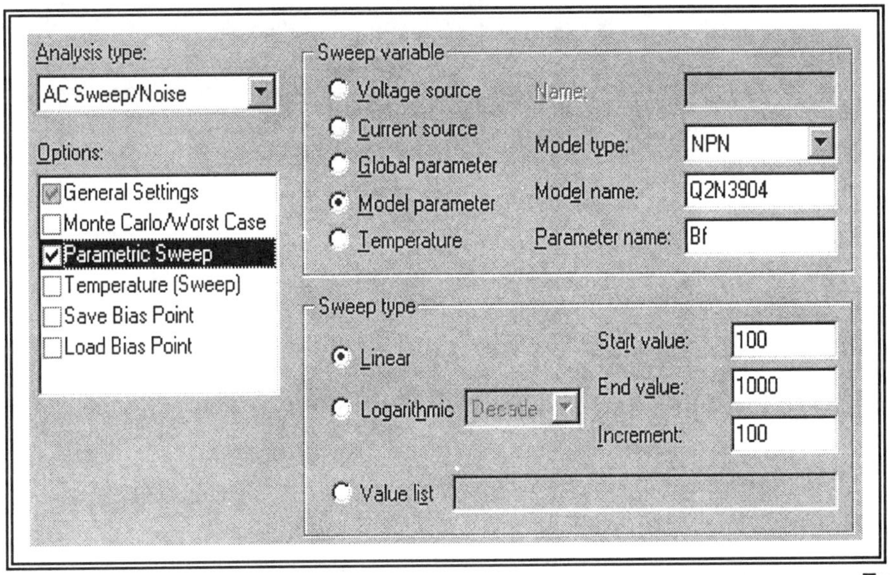

FIGURE 9.9

Adding a parametric setting

19. Run PSpice and generate the expanded Z_{IN} family of curves of Figure 9.10.

20. Based on the results of Figure 9.10, by what approximate percentage did Z_{in} increase when the *maximum Beta* increased from 100 to 1000?

        % change in Z_{IN} = _____

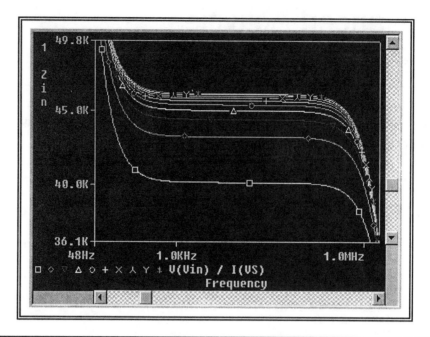

FIGURE 9.10

Family of Z_{IN} curves for
various values of Bf

Darlington Buffer

21. Add schematic DARLINGTON to project buffer, and draw the buffer circuit of Figure 9.11, which uses a Darlington pair to greatly increase its effective β ($\beta_{\text{total}} = \beta 1 \times \beta 2$).

22. Using AC analysis, measure midband Z_{IN} and Z_{OUT} for the Darlington buffer and compare to the non-Darlington values.

Darlington	**Non-Darlington**
Z_{IN} = _____	Z_{IN} = _____
Z_{OUT} = _____	Z_{OUT} = _____

23. Based on step 22, does a Darlington buffer increase the isolation properties of a buffer?

Yes   No

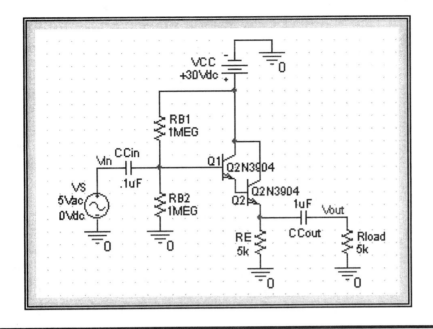

FIGURE 9.11

Darlington
buffer

24. For the Darlington pair buffer of Figure 9.11, determine the amount of quiescent (DC) power dissipated in each transistor (Q1 and Q2). Which transistor is likely to require a heat sink?

25. The low-frequency breakpoints shown in Figure 9.3 result from the two *RC* lead/lag networks involving capacitors *CC1* and *CC2*. By Thevenizing the input and output circuits, calculate (by hand) the two breakpoints and compare with Figure 9.3.

Exercises

1. Perform a complete analysis of the amplifier/buffer of Figure 9.12, including such items as A(overall), G, Z_{IN}, Z_{OUT}, bandwidth, distortion, power, and energy. (How does the buffer stage protect the gain of the amplifier stage?)

2. Using transient and parametric analysis, investigate the properties of the *buffered* voltage regulator of Figure 9.13. (How low can we take R_L before the system comes out of regulation?) Compare your answer to the results of Chapter 3.

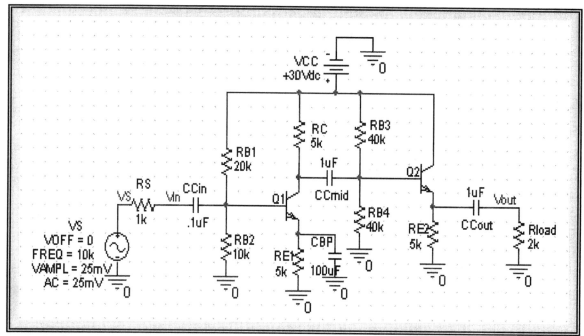

FIGURE 9.12

Amplifier/buffer

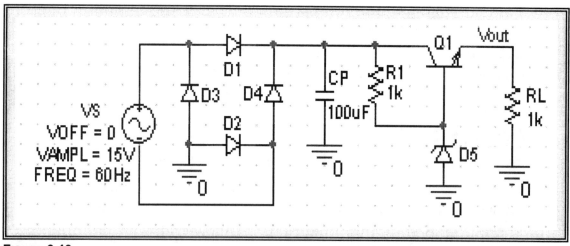

FIGURE 9.13

Buffered voltage
regulator

3. Referring to the buffer of Figure 9.1, replace *VAC* with *VSIN* and enter appropriate amplitude, offset, and frequency attributes. Select the transient mode, generate V_{OUT}, and determine the amplitude distortion from the following equation.

$$\% \ distortion = \frac{Vpeak(difference)}{Vpeak(average)} \times 100 = \underline{\hspace{2cm}}$$

Questions and Problems

1. In the following statement, fill in each blank with "collector" or "emitter":

   To achieve voltage gain, we tap the output voltage off the _____, and to achieve buffering action, we tap the output voltage off the _____.

2. Why are the buffer circuits of this experiment called *emitter followers*? (What is the relationship between input and output voltage amplitude and phase?)

3. For every 100,000 electrons that enter the emitter of a Darlington pair, how many electrons leave the base?   (Assume $\beta = 175$ for each transistor.)

4. Why are all the buffer circuits of this experiment highly linear?   Would they remain highly linear if the load were reduced to 25Ω?   Why?

5. By using the words "high," "low," or "medium" in the following spaces, contrast the amplifier of Figure 8.1 with the buffer of this chapter.

	Amplifier	Buffer
A	_____	_____
Z_{IN}	_____	_____
Z_{OUT}	_____	_____
Linearity	_____	_____

6. Consider a buffer with Zin and $Zout$. For maximum transfer of voltage from left to right (source to load), which of the following should exist?

    a. Z_{IN} low, Z_{OUT} low

    b. Z_{IN} low, Z_{OUT} high

    c. Z_{IN} high, Z_{OUT} low

    d. Z_{IN} high, Z_{OUT} high

7. Referring to Figure 9.14, why is the input impedance at the base approximately equal to 100kΩ?   Why is the output impedance at the emitter approximately equal to 10Ω? (Assume that re' is zero.)

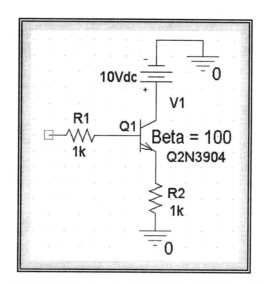

FIGURE 9.14

Simple buffer

8.   Referring to Figure 9.5, why does Z_{IN} go up at low frequencies and down at high frequencies?

9.   Contrast and compare the two methods of measuring Z_{OUT} introduced in Chapters 8 and 9.   (*Hint*: Refer to Figures 8.5 and 9.6.)

10

Amplifier Power

Class A Operation

Objectives

- *To analyze a Class A amplifier*
- *To generate a load line*
- *To generate a damped sine wave*
- *To determine power factors*

Discussion

Unlike the small-signal amplifier of Chapter 8, the Class A amplifier of Figure 10.1 is designed for large voltage and current applications (note the small resistor values). Therefore, power is an important consideration. Because we have access to a split power supply, we choose the simple and stable emitter biasing.

In particular, our design must provide the following:

- The maximum possible unclipped output voltage.
- The ability to handle the heat dissipated in the transistor.

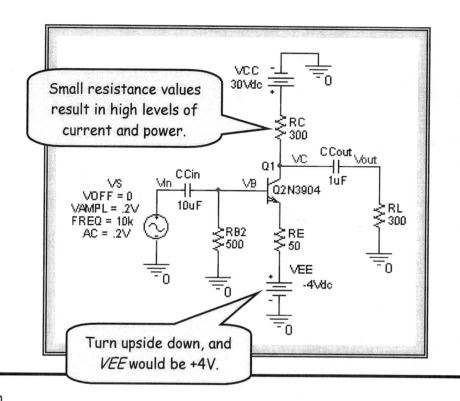

FIGURE 10.1

Initial Class A
amplifier design

Load Line

Our primary Class A design aid is the *load line*. As shown by Figure 10.2, we first locate the Q point, then we draw a line through the Q point whose inverse slope is equal to the total AC load.

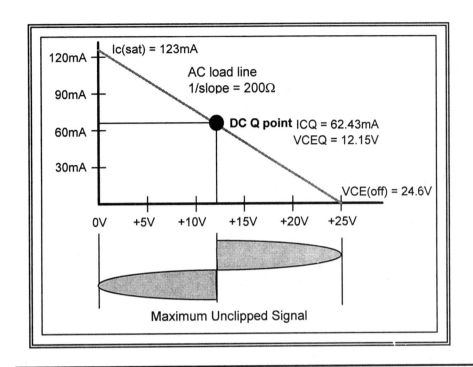

FIGURE 10.2

The load line

When an AC signal is present, the operating point moves back and forth along the load line about the Q *point*. The two endpoints are called $I_{C(SAT)}$ and $V_{CE(OFF)}$. If the circuit is well designed, the Q point is *centered* and we generate the largest possible output signal without clipping.

The values shown on Figure 10.2 were calculated as shown below. (The saturation and cutoff values are easily determined from the geometry of similar triangles.)

As before, we base all calculations on midband operation in which all capacitors are assumed to be dead shorts.

$I_{EQ} = I_{CQ} = (4V - .7V) / (50\Omega + .5k\Omega/175) = 62.43mA$

$V_{CEQ} = V_{CQ} - V_{EQ} = (30V - 62.43mA \times 300) - (-4V + 62.43mA \times 50) = 11.27V - (-.88V) = 12.15V$

$rl \text{ (ac load)} = R_C \| R_L + R_E = 300\Omega \| 300\Omega + 50\Omega = 200\Omega$

PSpice for Windows

Based on Figure 10.2, we predict a well-designed circuit (with a nearly centered Q point). We further predict a *compliance* (maximum unclipped signal) of approximately 12.15V. (However, this is not the compliance of the true output signal across R_L because R_E absorbs 25% of the load voltage. Therefore, we predict that the output voltage compliance across R_L is 75% of 12.15V, or 9.11V.)

Power Considerations

- The *load power* (P_L) is the average AC power developed across resistor R_L (the true load) when driven at its maximum unclipped level. The load power is the usable power. For a compliance of 9.11V, we obtain the following.

$$P_L = .5 \times V_{OUT}^2 / R_L = .5 \times 9.11^2 / 300\Omega = 138mW$$

- The *source power* (P_S) is the average power supplied by the DC power supplies. The source power is what we pay for. It is conveniently calculated using the average current.

$$P_S = (V_{CC}+V_{EE}) \times I_C(average) = (30V+4V) \times 62.4mA = 2.1W$$

- The *dissipated power* (P_D) is the worst-case average power deposited into the transistor. Because *AC* voltage and current are out of phase within the transistor, the worst-case dissipated power is the quiescent *DC* power.

$$P_D = V_{CEQ} \times I_{CQ} = 12.4V \times 62.43mA = 774mW$$

- The *efficiency* (η) is determined by dividing the maximum load power by the source power and converting the result to a percentage.

$$\eta = P_L/P_S \times 100 = 138mW/2.1W \times 100 = 6.6\%$$

In this chapter, we verify all of these predictions using PSpice. During our experimental activities, a major consideration will be: Should we use *transient* or *AC* analysis—or both?

Simulation Practice

Activity *AMPLIFIER*

This activity will analyze a class A amplifier, paying particular attention to the three types of power: load, source, and dissipated.

---

1. Create project *classa* and draw schematic *AMPLIFIER* of Figure 10.1.

2. To determine the maximum unclipped output signal, we will input a sine wave of gradually increasing magnitude and note when clipping occurs on the output. Before generating a reverse damped sine wave, first review *Simulation Note 10.1*.

3. **DCLICKL** part *VSIN* and enter –4000 in the *df* field of the Property Editor box.

Simulation Note 10.1
How do I generate a damped sine wave?

As first explained in Chapter 6 (Volume I), *VSIN* uses the following formula to generate a sine wave:

$$V_{OUT} = V_0 + V_{AMPL}\sin(2\pi f(t-td)+p/360)e^{-(t-td)df}$$

As an example, let's generate a damped sine wave that starts from 200mV and rises to approximately 10V in 10 cycles of a 10k waveform. For this case, *df* (damping factor) is calculated as follows

$$V_{OUT}/V_{AMPL} = 10V/.2V = 50 = e^{-(1ms \times df)}$$

$$\therefore df = -4000$$

4. Set the simulation profile to *Transient* from 0 to 1ms, with a ceiling of 1µs.

5. Generate the input/output waveforms of Figure 10.3.

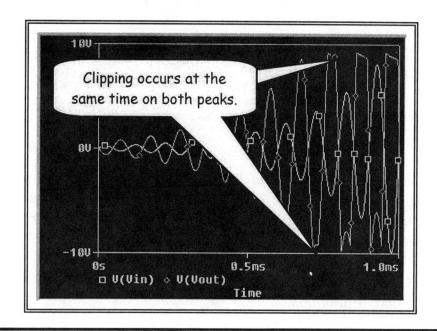

FIGURE 10.3

Looking for clipping

6. Does the input waveform (V_{IN}) rise from .2V to 10V in 10 cycles?

    Yes   No

7. Do the waveforms of Figure 10.3 verify that the circuit is well designed? (Does the output signal clip on both ends at approximately the same time, indicating the Q point is approximately centered?)

    Yes   No

8. Is the output signal compliance (maximum unclipped output signal) approximately 9V, as predicted?

    Yes   No

Power

9. Erase the amplitude waveforms of Figure 10.3, and use the power equations developed in the discussion to generate the *instantaneous* power waveforms of Figure 10.4. (For dissipated power, we can place the *W* marker directly on the transistor.)

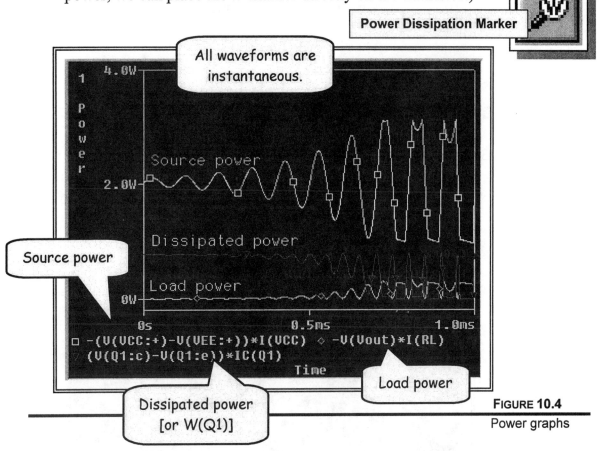

Power Dissipation Marker

All waveforms are instantaneous.

Source power

Dissipated power [or W(Q1)]

Load power

FIGURE 10.4
Power graphs

10. Based on the power curves of Figure 10.4, estimate each of the following. (Hint: Average *load* power at any point is one-half the peak power.)

- P_{LOAD} (average at max unclipped point) = _____

- $P_{DISSIPATED}$ (worst case) = _____

- P_{SOURCE} (average) = _____

- η (efficiency) = $P_{LOAD} / P_{SOURCE} \times 100$ = _____%

11. Do the experimental values of step 10 approximately equal the theoretical values calculated in the discussion?

    Yes   No

12. Using frequency domain analysis on the amplifier of Figure 10.1, generate a Bode plot and determine the voltage bandwidth.

    Bandwidth = _____

Advanced Activities

13. Generate the real-time load line of Figure 10.5. (*Hint*: Start with the same damped sine wave, but be sure to redefine the X-axis.) Can you tell approximately where the Q point is by the thickness and intensity of the curve? Compare the result to the predicted load line of Figure 10.2.

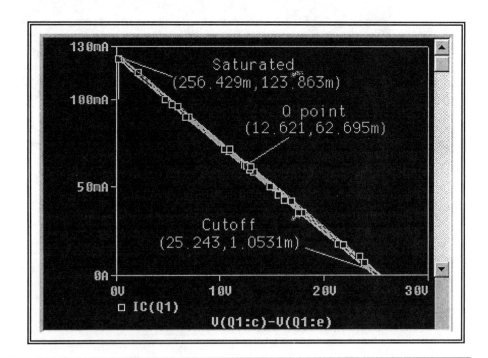

FIGURE 10.5

Load line created by PSpice

14. Compare the harmonic distortion of the output waveform before and after clipping occurs. (See Chapter 19, Volume I for a discussion of harmonic distortion.)

Exercises

1. Check the audio amplifier of Chapter 9 (Figure 9.12) for Class A operation. If necessary, make changes to bring it into Class A operation.

2. Change the Q point of the amplifier of Figure 10.1 to a non-centered condition and regenerate the load line of Figure 10.5.

Questions and Problems

1. How does a Class A amplifier differ from a small-signal amplifier?

2. If the Q point is centered, the output signal clips

    a. at the positive peaks first.
    b. at the negative peaks first
    c. at both positive and negative peaks at the same time.

3. Would you say that the efficiency of a Class A amplifier is high or low?

4. For a sine wave across the load resistor (R_L), the average current and voltage are zero.   Why is the average *power* not equal to zero?

5.   Why does the worst-case dissipated power (power lost in the transistor) occur when the *AC* signal input is zero?

6.   Why do *power* amplifiers use smaller resistors than *small-signal* amplifiers?

7.   Referring to the load line of Figure 10.5, why is the thickness and intensity of the load line greatest about the Q point?

8.   What is the difference between *power* and *energy*?

9.   Why is a load line straight (linear)?

10. If the load (R_L) is increased, would the load line slope of Figure 10.5 increase
     or decrease?

11

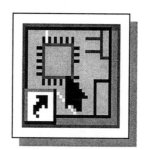

Amplifier Efficiency

Class B and C

Objectives

- *To analyze Class B and C amplifiers*
- *To compare the efficiency of Class A, B, and C amplifiers*

Discussion

The Class A amplifier of Chapter 10 suffers from notoriously poor efficiency. To greatly increase efficiency, we switch to the Class B and C amplifier/buffer designs of this chapter.

- The Class B circuit of Figure 11.1 employs two transistors in a push/pull configuration. Because we have the luxury of a split power supply, no coupling capacitors are required. The circuit is called Class B because each transistor is on (conducts) for approximately 50% of each cycle. The upper (NPN) transistor conducts during the positive half cycles and the lower (PNP) conducts during the negative half cycles.

  The Class B configuration is usually employed as a buffer and power amplifier in the common-collector (emitter-follower) configuration. A Class B buffer can yield more than 70% efficiency because very little power is wasted in biasing the circuit.

- The Class C circuit of Figure 11.2 is the most efficient of all. When in operation, it simulates the action of a hammer and a bell. A biased clipper is the hammer, and a tank circuit is the bell.

  At the top of each input cycle, the transistor saturates, the collector is grounded, and the capacitor is suddenly charged (the hammer hits the bell). The transistor then goes into cutoff for the rest of one or more cycles, and the capacitor and inductor trade energy (the bell rings). The amplifier is called Class C because the transistor is on for much less than 50% of each cycle.

  A Class C amplifier typically yields more than 90% efficiency because almost no power is lost in biasing the circuit or is dissipated in the transistor. Because of its high efficiency and the use of a resonant tank circuit, the Class C amplifier is typically used as a common-emitter radio-frequency amplifier for frequencies above 1MHz.

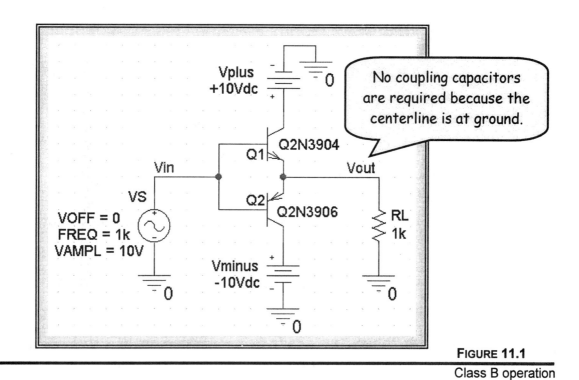

FIGURE **11.1**

Class B operation

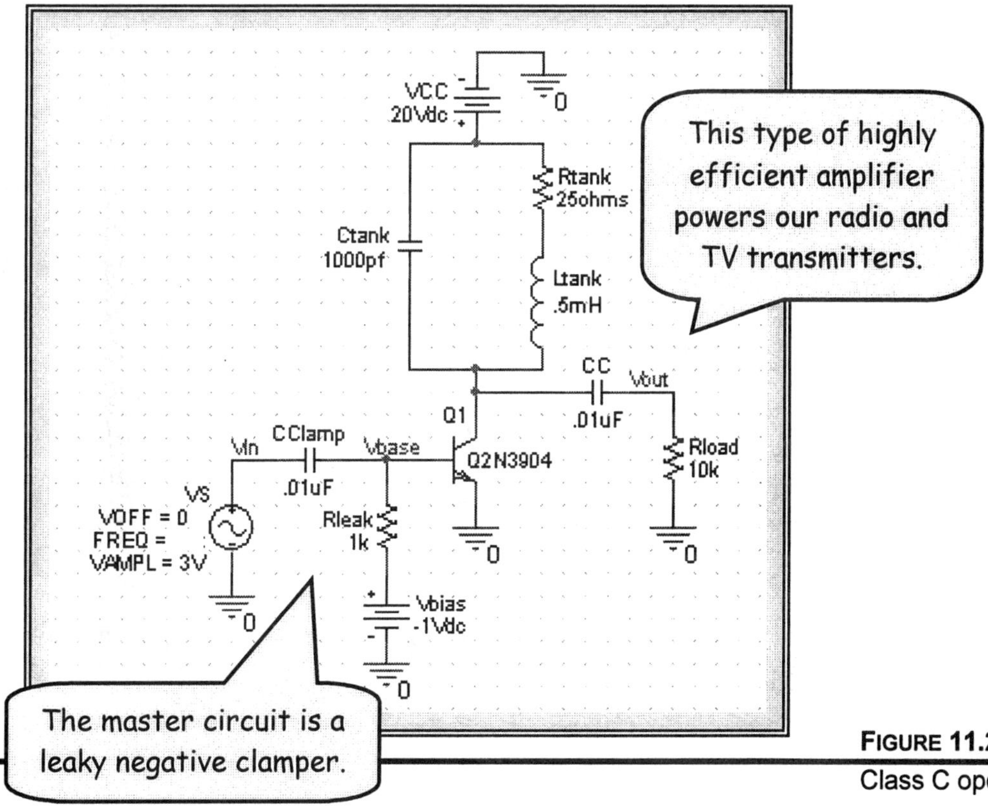

FIGURE **11.2**

Class C operation

Simulation Practice

Activity *CLASSB*

Activity *CLASSB* analyzes the Class B buffer of Figure 11.1, with special emphasis on power.

1. Create project *classbc* and schematic *CLASSB*, and draw the Class B buffer of Figure 11.1.

2. Set the simulation profile to *Transient* from 0 to 2ms, with a step ceiling of 2µs.

3. Use PSpice to generate the transient output waveform and determine the voltage gain.

    Class B voltage gain (A) = _____

    Is the voltage gain what you would expect of a buffer?

    Yes   No

4. Expand the output waveform (*Zoom Area* toolbar button) at the 1.5ms point where it crosses the 0V axis (Figure 11.3). This is called *cross-over distortion*, and it is caused by the barrier potential of the two transistors. (While the input signal is between +.7V and –.7V, the output signal is zero.)

5. To overcome cross-over distortion, add the *trickle-bias* circuit of Figure 11.4. The circuit makes use of the familiar barrier potential characteristics of diodes in order to bias each transistor just beyond its knee.

6. Generate new output waveforms. Is the cross-over distortion gone?

    Yes   No

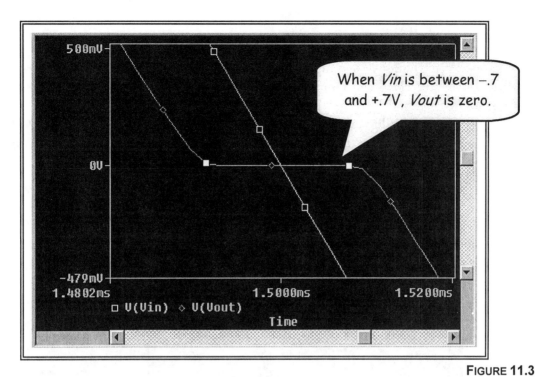

FIGURE **11.3**

Cross-over distortion

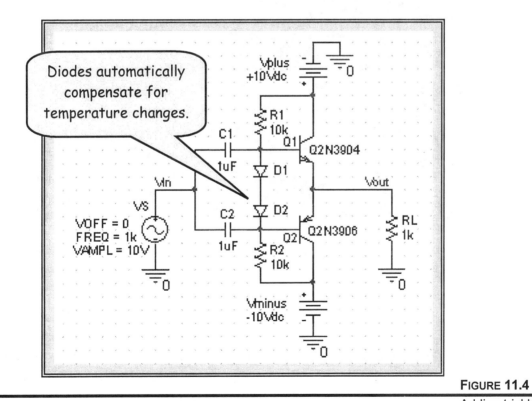

FIGURE **11.4**

Adding trickle bias

Efficiency

7. Efficiency is equal to *average* load power divided by *average* source power (reported as a percentage).

    For the trickle bias circuit of Figure 11.4, use the following midband equations to generate the *instantaneous* load and source power curves of Figure 11.5. (The minus signs are necessary because source and emitter currents are negative.)

    - $P(load) = -V(V_{OUT}) \times I(R_L)$

    - $P(source) = -(V(V+{:}+) \times I(V+) + V(V-{:}+) \times I(V-))$

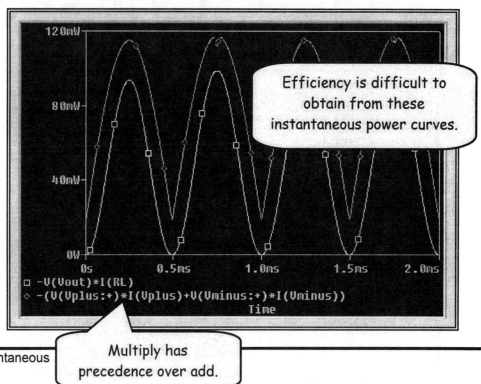

FIGURE 11.5

Class B instantaneous power curves

8. From the curves of Figure 11.5, we could obtain average power through mathematical analysis. Instead, use the *AVG* (average) operator to generate the running average curves of Figure 11.6.

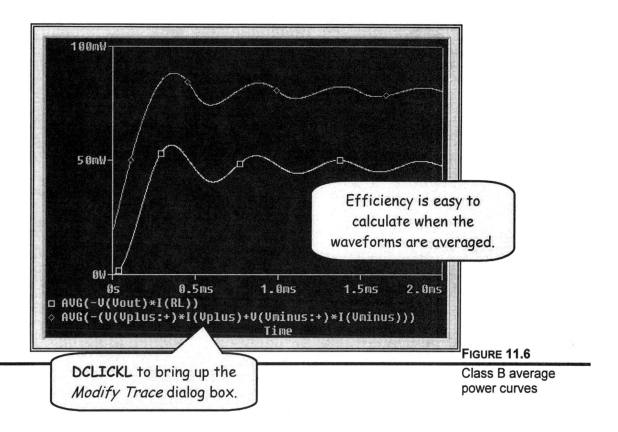

Efficiency is easy to calculate when the waveforms are averaged.

DCLICKL to bring up the *Modify Trace* dialog box.

FIGURE 11.6

Class B average power curves

9. Use your cursor to determine the steady-state average source and load power (the values near the right-hand side of the graph).

Average load power   = _____

Average source power = _____

10. From the results of step 9, determine experimental efficiency. (How does it compare with the maximum value mentioned in the discussion?)

$$\eta \text{ (efficiency)} = \frac{\text{Average load power}}{\text{Average source power}} \times 100 = \text{_____} \%$$

11. If we reduced the load, would the efficiency go up or down? Why?

Activity *CLASSC*

Activity *CLASSC* analyzes the Class C amplifier of Figure 11.2, with special emphasis on power and efficiency.

12. Add schematic *CLASSC* to project *classbc*, and draw the circuit of Figure 11.2.   (The frequency value of V_S will be set later.)

13. Determine the resonant frequency of the tank circuit.

$$F(\text{resonant}) = \frac{1}{2\pi\sqrt{LC}} = \underline{\hspace{2cm}}$$

14. Set the frequency value of V_S to approximately one-tenth of the resonant frequency.

15. Set the simulation profile to *Transient* from 0 to 150μs (step ceiling of 150ns), and generate the input, base voltage, and output waveforms of Figure 11.7.

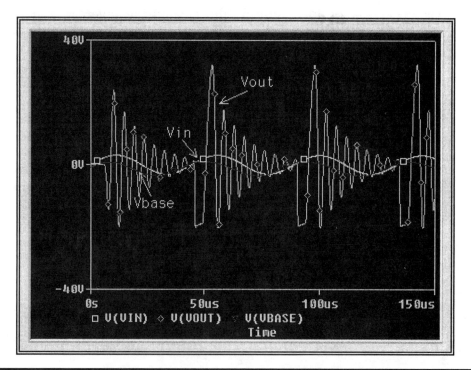

FIGURE 11.7

Class C input and output waveforms

PSpice for Windows

16. a. Is the base voltage negatively clamped?

    Yes   No

    b. Is the output waveform a damped sine wave at the resonant frequency?

    Yes   No

17. As time permits, change the value of the input frequency and note the result.

Advanced Activities

18. Perform an energy analysis of the Class C amplifier and *estimate* its efficiency.

19. Regarding the Class B buffer, compare its harmonic distortion before and after trickle bias.

Exercises

1. Using PSpice results, compare the overall efficiency of the audio amplifier of Figure 11.8 (using a Class B output stage) with the overall efficiency of the audio amplifier of Figure 9.12 (using a conventional output stage). Note that the amplifier uses a single power supply and is able to drive an 8Ω speaker.

2. Increase the Q (lower *Rtank*) of the Class C amplifier of Figure 11.2 and compare the results to the high-Q circuit.

Questions and Problems

1. The input circuit to the base of a Class C amplifier is

   a. a positive clamper.
   b. a negative clamper.

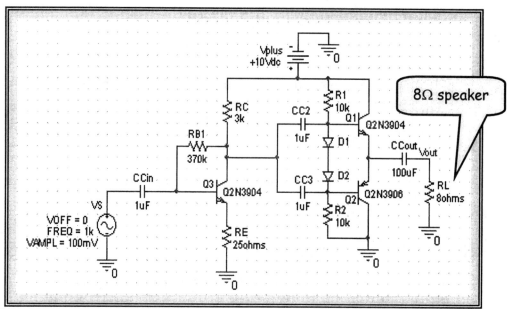

FIGURE 11.8

Audio amplifier

2. Referring to the biased Class B buffer stage of Figure 11.8, what provides power to the circuit during the half cycle when the upper NPN transistor is biased off? (*Hint*: What circuit component stores energy?)

3. Why does the trickle-bias circuit of Figure 11.4 consume very little power?

4.   Assuming that the transistors of Figure 11.1 have a *Beta* of 175, what is the approximate power gain?

5.   Viewing the Class C waveforms of Figure 11.7, what causes the output waveform to decay between hits?

6.   Why is the efficiency of a Class C amplifier so high?   (*Hint*: Why does most of the source energy pass on to *RL*?)

7.   For each of the amplifier classes listed below, approximately what percentage of the time is a given transistor on?

   Class A _____

   Class B _____

   Class C _____

8.   Why is a Class C amplifier used to power the final stage of a radio or television transmitter?

9.   Why is the trickle bias circuit of Figure 11.4 often called a *current mirror*?

Part 3

Field-Effect Transistor Circuits

In Part 3, we move from the bipolar to the field-effect transistor (FET). As in Part 2, we concentrate primarily on amplifiers and buffers.

We will see how the FET's inherent high input impedance and vastly different transconductance properties influence its characteristics.

Of special note is the CMOS configuration, the basic switching element of digital integrated circuits.

12

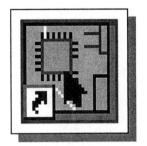

The Field-Effect Transistor

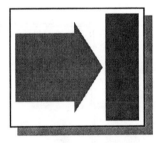

Drain Curves

Objectives

- *To display FET master and slave curves*
- *To determine an FET's input and output impedance*
- *To determine the effects of temperature on an FET*

Discussion

Like the bipolar transistor, a *field-effect transistor* (FET) is also a *voltage-controlled current source* (VCIS). That is, an input voltage controls an output current, regardless of the output voltage. However, the way the FET achieves its VCIS characteristics is vastly different.

The FET comes in a number of different types. The JFET (*junction field-effect transistor*) of Figure 12.1 was developed first. It achieves its VCIS characteristics as follows: As the positive drain/source voltage increases, it pulls on electrons, creating a drain current in the N region. However, because of the reverse-biased PN junction between gate and drain, it also creates a positively charged *depletion region* that squeezes the drain current.

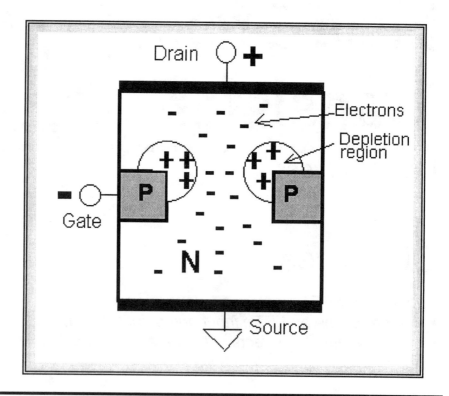

FIGURE 12.1

JFET internal
operation

In normal operation, when the *drain voltage* is above a certain threshold (*pinchoff*), this pulling and squeezing cancel, *and the drain becomes a current source* (current independent of voltage). However, the *input* gate/source voltage is not subject to cancellation (it only squeezes) and therefore *does* control the current.

In summary, the input gate/source (master) voltage controls the output drain (slave) current, regardless of the drain/source voltage, making the FET a *voltage-controlled current source* (VCIS).

Bipolar Versus JFET

If both the bipolar transistor and FET are simply VCISs, then how do they differ? The most important differences are the JFET's extremely high input impedance and its vastly different input/output (transconductance) characteristics. These two differences give the JFET an edge when used in certain applications.

Test Circuit

The test circuit of Figure 12.2 shows an N-channel JFET. The JFET also comes in the P-channel version but may not be available in the PSpice evaluation library. The P-channel is identical in operation to the N-channel, except all voltages and currents are reversed.

Note the complete lack of resistance in the master (gate/source) circuit. Since the JFET is a purely voltage-controlled device, essentially no current at all flows in the master circuit.

Transconductance

When operated as a VCIS, the output of an FET is drain current (I_D), and the input is gate/source voltage (V_{GS}). This ratio ($\Delta I_D / \Delta V_{DS}$) is called the transistor's *transconductance* (g_M) and is given in units of μSeimens (μS). A typical value for transconductance is 4000μS (or $1/250 \, \Omega$).

As before, one of the best ways to investigate the VCIS characteristics of a JFET is by generating drain and gate/source curves. This is done with the test circuit of Figure 12.2.

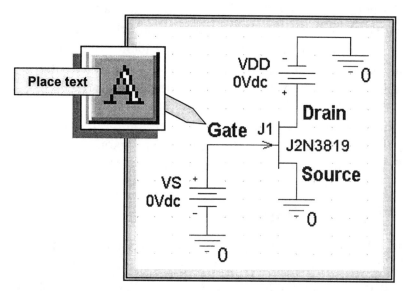

FIGURE 12.2
N-channel JFET
test circuit

Simulation Practice

Activity *JFET*

Activity JFET will generate drain and transconductance curves for the J2N3819 JFET of Figure 12.2.

1. Create project *fet* (with schematic *JFET*), and draw the circuit of Figure 12.2. (V_S and V_{DD} will be swept and therefore need not be assigned bias point values at this time.)

2. Set the simulation profile to DC Sweep, with a primary sweep of *VDD* from 0 to 10V (increment .1V) and a secondary (nested) sweep of *VS* from 0 to −3V (increment 1V).

Drain Curves

3. Run PSpice and generate the drain curves of Figure 12.3.

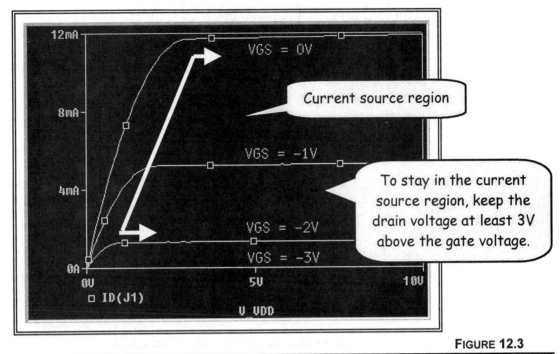

FIGURE 12.3

JFET drain
curves

4. By taking measurements, answer the following:

   a. What is the maximum possible current (known as I_{DSS})?

      $I_{DSS} = $ _____

   b. What input voltage (V_{GS}) produces I_{DSS}?

      V_{GS} at $I_{DSS} = $ _____

   c. What gate/source voltage causes *pinchoff* ($I_D = 0$)?   This
      value is called V_{GSOFF}.

      $V_{GSOFF} = $ _____

   d. Determine a typical value for output (drain) AC impedance
      in the current-source region.   (*Hint*: Measure 1/slope or use
      the "d" operator.)

      $Z_{OUT} = $ _____

Gate/Source (Master) Curve

5. By taking values from Figure 12.3 (in the current-source region), draw by hand a *transconductance* curve on the graph of Figure 12.4.

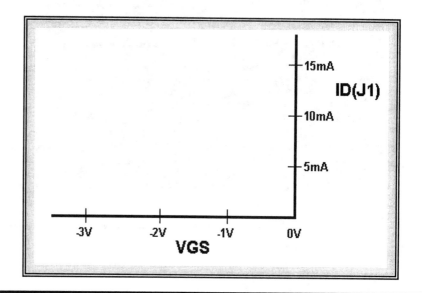

FIGURE 12.4

Transconductance
curve

6. Use PSpice to generate the transconductance curve of Figure 12.5 *directly*, and compare to the *indirect* method of Figure 12.4.   (*Hint*: Set V_{DD} to +10V and do a DC Sweep of V_S.)

7. The master curve of Figure 12.5 is a *parabola* that obeys the following equation:

$$I_D = I_{DSS} (1 - V_{GS}/V_{GSOFF})^2$$

As an example of the use of this equation, at what V_{GS} does I_D = 1/2 I_{DSS} (a good Q point location)? Does your value agree with Figure 12.5? (Obtain I_{DSS} and V_{GSOFF} from step 4.)

$$V_{GS} \text{ (at 1/2 } I_{DSS}) = \underline{\hspace{2cm}}$$

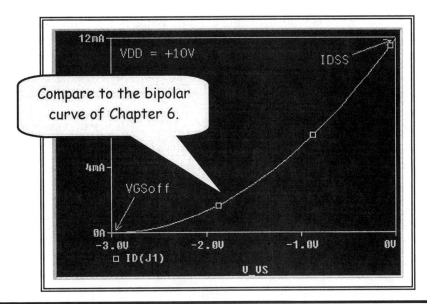

FIGURE 12.5

JFET transconductance
curve

8. Transconductance is the *slope* of the curve *($\Delta I_D/\Delta V_{GS}$)*. Using the differentiate ("d") operator, add a curve of transconductance to the graph of Figure 12.5. (The result is shown in Figure 12.6.)

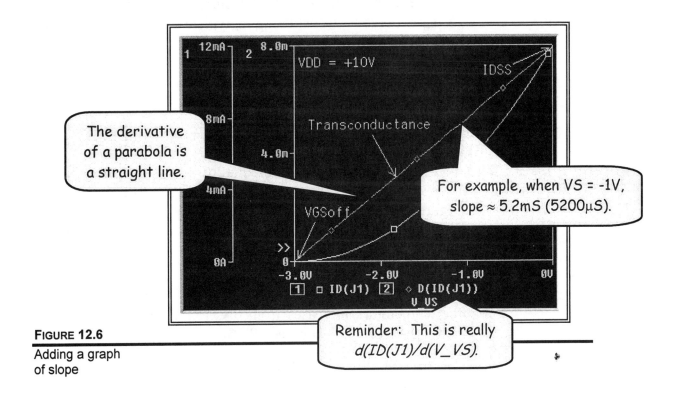

FIGURE 12.6

Adding a graph
of slope

9. Based on Figure 12.6, what is the transconductance at the half-current Q point (where $I_D = 1/2\ I_{DSS}$)?

    Transconductance at half-current = _____ μS

10. Add a third Y-axis to the graph of Figure 12.6, and plot input impedance $(V(J1{:}g)/IG(J1))$.   Record below the value for Z_{IN} at the half-current Q point.   (*Note*: "T" = *tera* = 10^{+12}.)

    Z_{IN} (Q point) = _____

    a. Based on your results, would you say that a JFET has naturally high input impedance?

        Yes   No

    b. Does Z_{IN} drop (but still remain high) as *VG* falls (from –3V to close to 0V)?

        Yes   No

    c. Delete the transconductance curve and substitute a plot of gate current $(IG(J1))$. Is the current generally in the Pico amp range?   Considering a typical value of Z_{IN}, is this reasonable?

        Yes   No

Master Curve Temperature Effects

11. To determine how temperature affects the master (transconductance) curve of Figure 12.5, recreate the graph of Figure 12.7. (Keep the primary sweep of *VS*, and make temperature a nested DC Sweep from –50°C to +50°C in increments of 25°C.)

12. Based on the results of step 11, answer the following: When the temperature increases from 0°C (32°F) to +25°C (103°F), IDSS changes by what percent?

    % change in I_{DSS} = _____

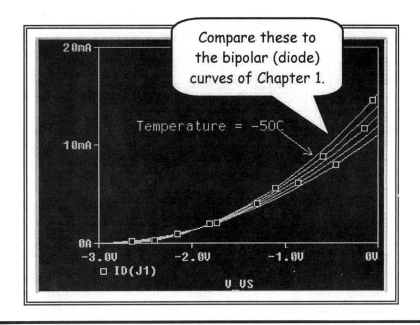

FIGURE 12.7

JFET temperature effects

Advanced Activities

13. Referring to Figure 12.3, we see that an FET's ohmic region (to the left of the current-source region) acts as a *voltage-variable resistor* for small voltages. Using the test circuit of Figure 12.8, generate the curves of Figure 12.9.

14. Based on the results of Figure 12.9, complete the following table, showing how *Vcontrol* changes the ohmic value of the JFET.

Vcontrol	Ohmic resistance
–2.4V	
–2.8V	
–2.9V	
–2.94V	
–2.96V	

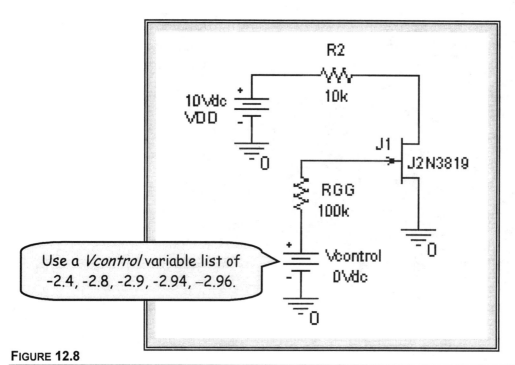

FIGURE 12.8

Voltage-variable
test circuit

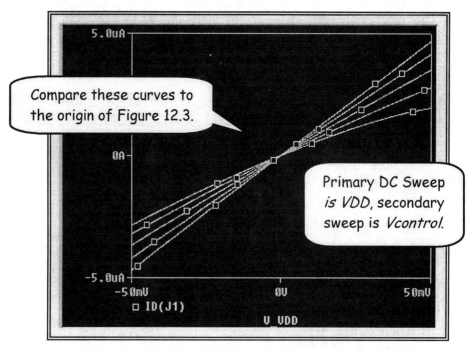

FIGURE 12.9

Variable resistance
curves

Exercises

1. Use master (gate source) and slave (drain) curves to plot the characteristics of the J2N4393 JFET (in library *EVAL*). How does the J2N4393 differ from the J2N3819?

2. Turn the J2N3819 JFET of Figure 12.2 upside down and again plot master and slave curves.   Summarize your results.

Questions and Problems

1.   Why is the input impedance of a JFET so high?

2.   Why is a JFET (when biased as in Figure 12.1) a VCIS (rather than an ICIS)?

3. Which of the following typically has the highest transconductance (output current change divided by input voltage change)?

   a. Bipolar transistor
   b. JFET

4.   Write "bipolar" or "JFET" before each statement.

   _____ Normally off, requires voltage to turn on.

   _____ Normally on, requires voltage to turn off.

5. The transconductance of a JFET is highest

    a. at low values of drain current.
    b. at high values of drain current.

6. A transconductance of 4000µS is equivalent to what in inverse ohms?

13

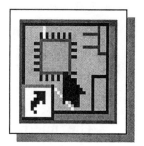

The
E-MOSFET

CMOS

Objectives

- *To display E-MOSFET master and slave curves*
- *To determine the characteristics of CMOS*

Discussion

Field-effect transistors fall into two general categories: *Depletion* and *enhancement*. The JFET of the last chapter is purely a depletion device. The E-MOSFET of this chapter is purely an enhancement device.

The internal construction of the E-MOSFET of Figure 13.1 reveals an NPN configuration. In order for electrons to travel from source to drain, the gate must build a bridge of electrons across the P region. This is accomplished by placing a positive threshold voltage on the gate to enhance the P region. Once this bridge of electrons is built, any further increase on the drain voltage pulls on electrons but also pinches the bridge. This cancellation effect results in a VCIS action similar to that of the JFET.

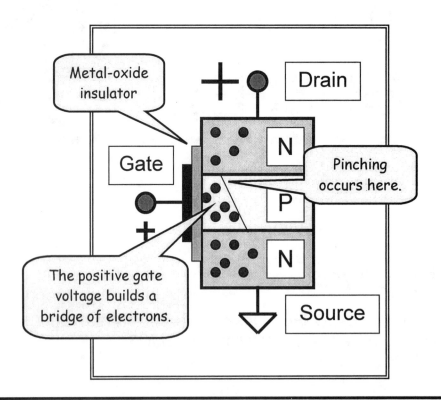

FIGURE 13.1

E-MOSFET internal
operation

Test Circuit

The test circuit of Figure 13.2 will generate drain and gate/source curves for the N-channel E-MOSFET. Equally important is the P-channel E-MOSFET, which is identical in operation to the N-channel, except all voltages and currents are reversed.

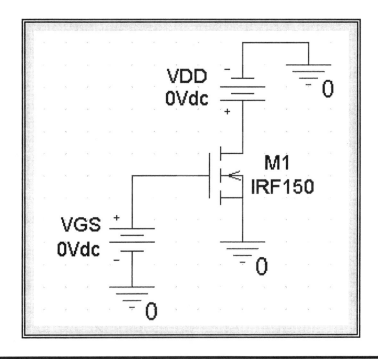

FIGURE 13.2

N-channel E-MOSFET test circuit

CMOS

The E-MOSFET has achieved stardom because its characteristics are perfectly matched to the needs of digital circuitry. In particular, when we stack a P-channel on top of an N-channel in a totem pole arrangement, we create the amazing *CMOS (complementary-symmetry metal-oxide semiconductor)* configuration of Figure 13.3.

The CMOS configuration is the basic digital switching unit of large-scale integrated circuits. As we will see, their widespread use results from their extremely low power consumption. Many variations of this basic configuration are in use today.

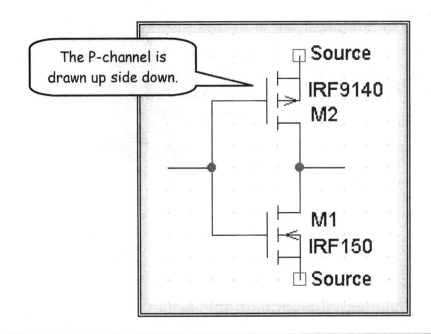

FIGURE 13.3

CMOS configuration

Simulation Practice

Activity *NCHANNEL*

Activity *NCHANNEL* will generate drain and transconductance curves for the IRF150 N-channel MOSFET of Figure 13.2.

1. Create project *emosfet*, with schematic *NCHANNEL*, and draw the circuit of Figure 13.2.

2. Set the simulation profile to *DC Sweep*, with a primary sweep of *VDD* from 0 to 10V in increments of .1V and a secondary (nested) sweep of VGS from 3V to 6V in increments of 1V.

Drain Curves

3. Run PSpice and generate the drain curves of Figure 13.4.

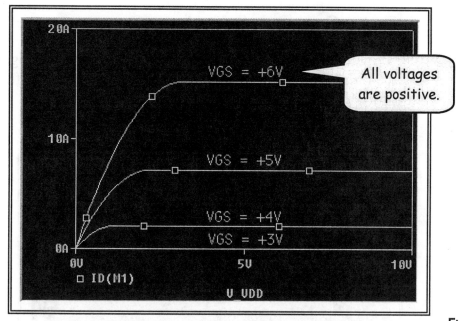

FIGURE 13.4

E-MOSFET drain
curves

4. Are all voltages positive, and does slave current (*ID*) increase as
    master voltage (*VS*) increases?

                    Yes   No

5. By taking measurements, answer the following:

    a. The E-MOSFET turns off for all gate/source voltages below
        what value? (This value is called V_{GSOFF}.)

        V_{GSOFF} = _____

    b. What is a typical value for output (drain) AC impedance in
        the current source region? (*Hint*: Measure 1/slope or use the
        "d" operator.)

        Z_{OUT} = _____

6. In order to be in the current source region, the drain voltage
    (*VD*) must be approximately three volts higher than the master
    voltage (*VGS*).

                    Yes   No

The Gate/Source (Master) Curve

7. Use PSpice to generate the transconductance curve of Figure 13.5. (*Hint*: Set V_{DD} to +10V and do a DC Sweep of V_S from +3V to +6V in increments of .1V.)

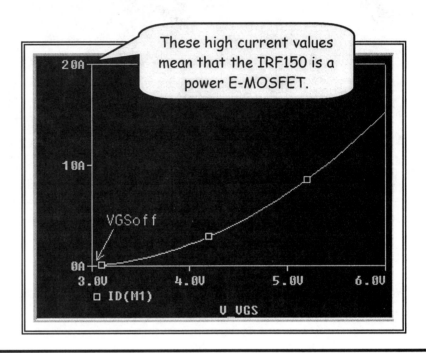

FIGURE 13.5

JFET transconductance curve

8. Based on the results of Figure 13.5, is the curve a classic FET parabola, and is VGS_{OFF} +3V, as predicted by the drain curves?

    Yes   No

Activity *CMOS*

The inverter of Figure 13.6 will be used to demonstrate the characteristics of CMOS. Of special importance is the low power consumption of CMOS.

9. Add schematic *CMOS* to project *emosfet*, and draw the circuit of Figure 13.6.

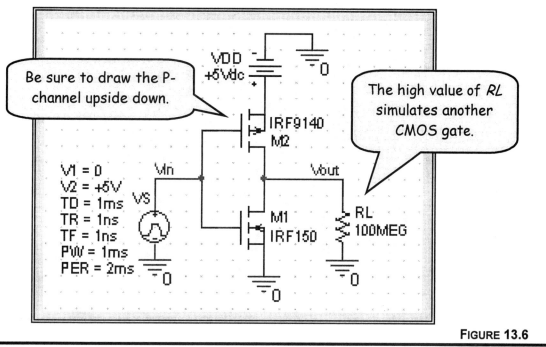

FIGURE **13.6**

Initial circuit

10. Set the simulation profile to *Transient* from 0 to 5ms, with a step ceiling of 5μs.

11. Generate the typical input/output waveforms of Figure 13.7.

    a. Is the output inverted from the input?

        Yes   No

    b. Is the output swing a full 5V (0V to 5V)?

        Yes   No

    c. Does a positive pulse appear at each positive-going switch?

        Yes   No

12. To perform a power/energy analysis of the CMOS circuit, use the *ABS* (absolute value) operator to generate the source power waveforms of Figure 13.8 (top plot).   Then use the *s* (integrate) operator to plot the energy graph of the bottom plot.

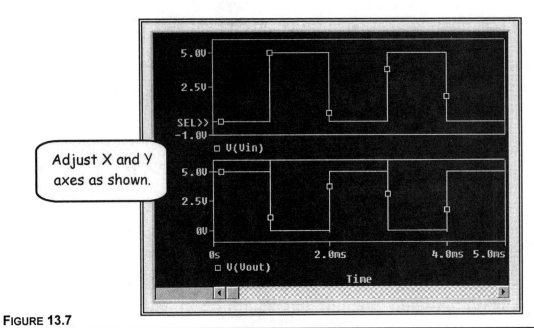

FIGURE 13.7

The CMOS inverter
waveforms

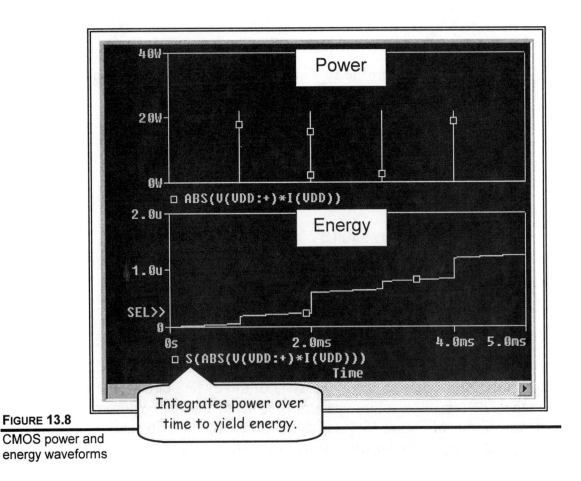

FIGURE 13.8

CMOS power and
energy waveforms

13. Reviewing the results (Figure 13.8):

    a. Is it safe to say that the CMOS configuration consumes very little power?

       Yes   No

    b. Does source current flow (and is power consumed) only during the brief periods of switching?

       Yes   No

    c. After four complete cycles, how many joules of energy have been consumed?   (1 joule = 1 watt × 1 second.)

       Energy in four cycles = _____ joules

Advanced Activities

14. Referring to Figure 13.2, extend the sweep voltages of both *VDD* and *VGS* to larger and larger upper-limit values. Does the IRS150 E-MOSFET continue to exhibit VCIS properties? Does it break down?

15. Does the E-MOSFET exhibit "ohmic" effects about the origin of the drain curves? (*Hint*: See Chapter 12.)

Exercises

1. Draw a master temperature effects curve for the IRS150 E-MOSFET (similar to Figure 12.7 for the JFET).

2. Based on Figure 13.5, customize the test circuit of Figure 13.2 to yield a drain current of exactly 10A.

3. Draw the E-MOSFET circuit of Figure 13.9, and display *Vin1*, *Vin2*, and *Vout*. Based on the results, what is the purpose of the circuit? (*Hint*: It's a basic digital gate.)

4. Redesign the digital gate of Figure 13.9 for NOR operation.

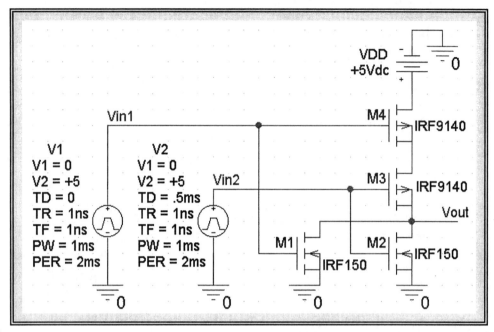

FIGURE **13.9**

Mystery CMOS
digital gate

Questions and Problems

1. What do the NPN transistor and E-MOSFET have in common?   (*Hint*: Compare the master curves of each.)

2. What does the first "E" stand for in the term *E-MOSFET*?

3.   Which term below would best characterize the Z_{IN} of an E-MOSFET?

   Very low   Low   High   Very high

4.   Why are E-MOSFETs especially useful in digital circuits?

5.   The NPN device of Figure 13.1 is like a canyon (the P-channel) surrounded on top and bottom by a high plateau (the N-channels). How does an electron then travel from source to drain?

6.   *Metal oxide* (a form of rust) is an

   Insulator   Conductor

7.   Referring to the CMOS configuration of Figure 13.3, why is the P-channel device drawn upside down?

8.   Based on the curves of Figure 13.4, estimate the value of ID when V_{GS} is raised to +7V.

9.　Based on the curves of Figure 13.7, what is an inverter?

10.　Why does the CMOS configuration use such a small amount of power?

11.　Develop a parabolic "master" curve equation for the transconductance curve of Figure 13.5. (*Hint*: Start with the general equation for a parabola, $y = a + bx^2$, and substitute values to determine a and b.)

14

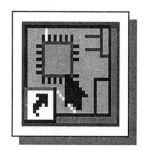

FET
Biasing

Temperature Stability

Objectives

- *To design and analyze FET biasing circuits*
- *To determine how temperature affects stability*

Discussion

We bias an FET for the same reason we bias a bipolar transistor: to place its quiescent (Q) point at an appropriate place in the master curve.

Based on the JFET master graph of Figure 14.1 (reproduced from Chapter 12), biasing requires a negative DC voltage across its gate/source junction. When properly biased, the superimposed AC signal will have room to operate on both its positive and negative cycles. Due to the parabolic shape of the transconductance curve, we can expect a fair amount of harmonic distortion in the output waveform.

As with the bipolar transistor, we can choose from a variety of JFET biasing circuits. The major considerations are simplicity, stability, and flexibility.

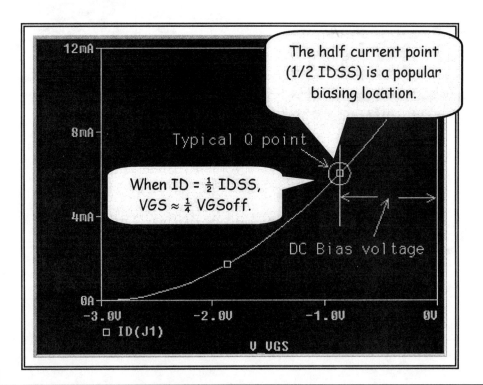

FIGURE 14.1

JFET master curve
bias (Q) point

Simulation Practice

Activity *SELFBIAS*

Activity *SELFBIAS* uses the circuit of Figure 14.2 to analyze this simplest type of JFET biasing. (We use the word *self* because rising current through *RS* places a positive bias voltage at the source, while the gate voltage remains at zero volts.)

Hand Calculations

1. By again solving the equations listed on the following page (as we did in Chapter 12), determine the gate/source quiescent voltage (V_{GS}) that places the Q point at the half-current biasing point (when $I_D = 1/2\ I_{DSS}$).

    $I_{DSS} = 12mA$ (from Figure 14.1)

    $V_{GSOFF} = -3V$ (from Figure 14.1)

    $I_{DQ} = I_{DSS}\ (1 - V_{GS}/V_{GSOFF})^2$

    V_{GSQ} (for half-current biasing) = _____ V

2. Is the half-current value for V_{GS} calculated in step 1 close to that indicated by Figure 14.1?   (If not, redo your calculations.)

          Yes   No

3. From the value of V_{GS} (at half-current) determined in step 1, use Ohm's law to calculate the value of R_S required to give this half-current Q point.   (*Hint*: V_G is approximately zero.)

    R_S (half-current bias point) = _____

PSpice Analysis

4. Create project *fetbias* with schematic *SELFBIAS*, and draw the circuit of Figure 14.2.   Be sure to set R_S to the value determined in step 3 ($\approx 150\Omega$).

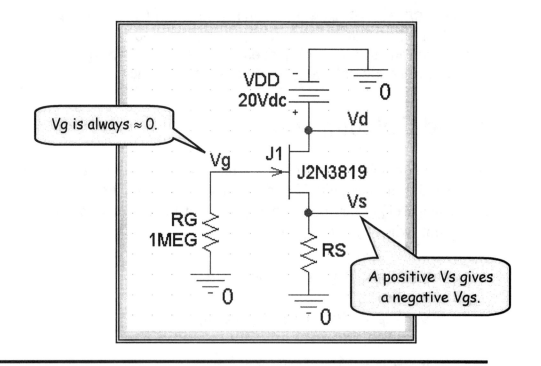

FIGURE 14.2

Self-biasing

5. From step 1, fill in the calculated Q point values. Using PSpice, perform a bias-point analysis and record the actual (PSpice) Q point values below.   Are the actual values approximately equal to the calculated values?

	Calculated	PSpice
I_{DQ}		
V_{GSQ}		

Temperature Effects

6. To determine temperature stability, we will see how the drain current [ID(J1)] changes with temperature. Generate the graph of Figure 14.3 and determine the following:

    $\Delta I_D / \Delta T = $ _____

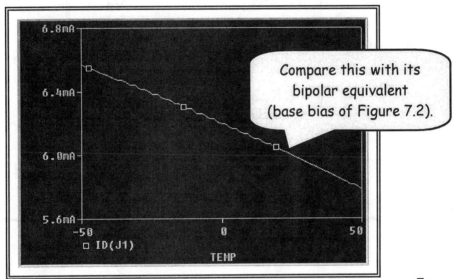

Compare this with its bipolar equivalent (base bias of Figure 7.2).

FIGURE 14.3

Self-bias temperature stability

Activity *VDBIAS*

Activity *VDBIAS* shows how voltage-divider bias is used to improve stability.

Hand Calculations

7. Use Ohm's law to determine the value of R_S required to place the Q point at the half-current point. (*Hint*: *Vsource* will be more positive than *Vgate* by the half-current DC bias voltage.)

    R_S (half-current point) = _____

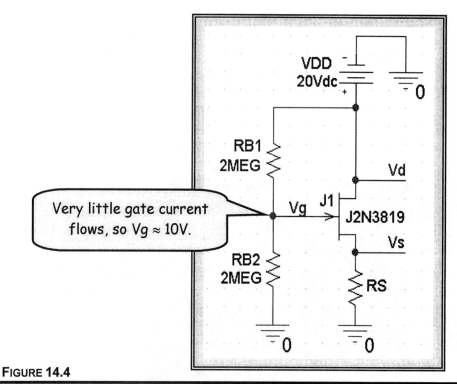

FIGURE 14.4

Voltage-divider bias

PSpice Analysis

8. Add schematic *VDBIAS* to project *fetbias*, and draw the circuit of Figure 14.4.   (Set R_S to the value determined in step 7.)

9. Use PSpice to perform a bias-point analysis and record the Q point values.   Are the actual values approximately equal to the expected values?   (Hint: RS ≈ 1.8k.)

	Calculated	PSpice
I_{DQ}	6mA	
V_{GSQ}	.879V	

10. Perform a temperature analysis of voltage-divider bias (similar to step 6) and create the comparison graph of Figure 14.5. (To append the *self-bias* curve, **File**, **Append Waveform**, select file, **Do Not Skip Sections**.)

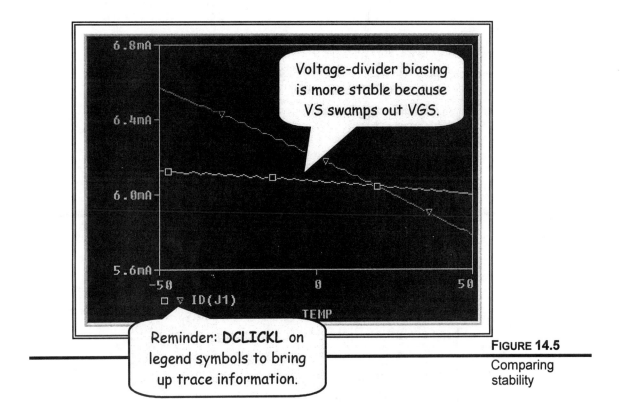

FIGURE 14.5
Comparing stability

11. Based on Figure 14.5, by what factor is voltage-divider biasing more stable than self-biasing? (Hint: Compare the slopes.)

    VDbias is _____ times more stable than *selfbias*.

Advanced Activities

12. Based on the graphical data of the previous chapter (Figure 13.5), determine the value of R_D that will give the E-MOSFET drain-feedback biasing circuit of Figure 14.6 a bias current of approximately 4A. Check your results using PSpice. (*Hint*: Assume that $I_{RG} = 0$.)

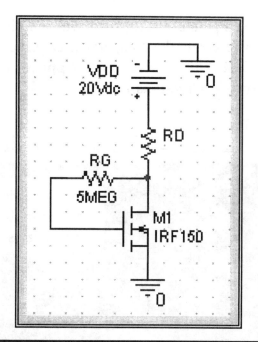

FIGURE 14.6

Drain feedback
biasing of
E-MOSFET

13. Perform a temperature sensitivity analysis of the E-MOSFET bias circuit of Figure 14.6.

14. For each of the bias circuits of this chapter, determine the power dissipated in the transistor.

Exercises

1. Perform a sensitivity analysis on the voltage-divider bias circuit of Figure 14.4. The Q point is most sensitive to which resistor? (If necessary, see Chapter 7.)

2. Based on Figure 14.2, design a self-bias circuit with an I_{DQ} of 8mA.

3. Determine the temperature stability of the E-MOSFET bias circuit of Figure 14.6. Compare to the JFET bias circuits.

Questions and Problems

1. Referring to Figure 14.1, when the current is at the half-current point (6mA), the voltage is *approximately* at the

   (a) quarter-voltage point (–.75V).
   (b) half-voltage point (–1.5V).

2. Why is the biasing technique of Figure 14.2 called "self-biasing"?

3. Based on the transconductance curve of Figure 14.1, what gate voltage will place the JFET in Figure 14.7 at the half-current point?

   V_G (half-current) = _____

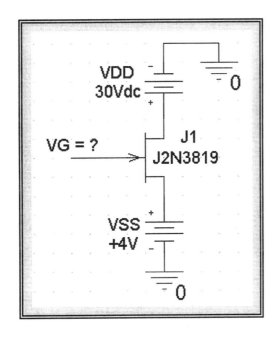

FIGURE 14.7

JFET biasing
test circuit

4. Referring to the voltage-divider biasing circuit of Figure 14.4, why can we be quite sure that the gate voltage is very close to 10V?

5. Can we use the drain feedback biasing method of Figure 14.6 on a JFET? Why or why not?   (*Hint*: With a JFET, can V_G ever equal V_D?)

CHAPTER

15

FET
Amplifiers
and Buffers

Input Impedance

Objectives

- *To design and analyze JFET amplifiers and buffers*
- *To determine FET circuit input and output impedance*

PSpice for Windows

Discussion

Amplifiers and buffers made with JFETs are similar in design and function to those made with bipolar transistors. The major difference is that FET circuits offer much higher input impedance, and the differences in transconductance characteristics lead to changes in gain and linearity.

Simulation Practice

Activity *AMPLIFIER*

Activity *AMPLIFIER* will determine the characteristics of the typical FET amplifier of Figure 15.1.

1. Create project *fetampbuffer* with schematic *AMPLIFIER*, and draw the JFET amplifier of Figure 15.1. (Note that the same half-current self-biasing circuit analyzed in the previous chapter biases the amplifier.)

Q Point

2. Use PSpice to run a bias point solution. Examine the output file and mark the Q point on the transconductance curve of Figure 15.2. Is the Q point at the expected half-current point? (Figure 15.2 is a copy of Figure 12.6 and need not be generated by the student.)

Yes   No

3. *Transconductance* is the slope of the transconductance curve. Based on Figure 15.2, what is the transconductance (g_m) at the Q point? (Compare your answer with step 9, Chapter 12.)

g_m (at Q point) = _____ µSiemens

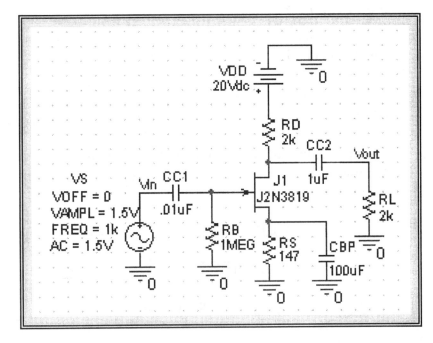

FIGURE 15.1

JFET amplifier using self-biasing

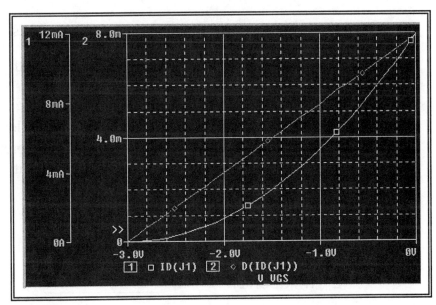

FIGURE 15.2

JFET transconductance curves

Hand Calculations

4. From Chapter 12, record typical values for each of the following:

$Z_{IN}(FET)$ = _____ $Z_{OUT}(FET)$ = _____

5. Using the value of g_m from step 4, calculate by hand the following:

A	=	$R_L \| R_D \times g_m$	=	_____
Z_{IN}	=	$R_B \| Z_{IN}(FET)$	=	_____
Z_{OUT}	=	$R_D \| Z_{OUT}(FET)$	=	_____

PSpice Analysis

6. Add the following simulation profiles to schematic AMPLIFIER

   - A *Transient* profile from 0 to 2ms, with a ceiling of 2µs.

   - An *AC Sweep* profile from 10Hz to 10GHz, with 100 points/decade.

7. Using either the transient or AC mode, determine the following items with PSpice and compare these results to the theoretical results of step 5. (To determine Z_{OUT}, measure V_{OUT} with and without a load.) Are they nearly the same?

A (midband)	=	_____
Z_{IN} (midband)	=	_____
Z_{OUT} (midband)	=	_____

Yes   No

8. Using *AC Sweep* mode, plot amplifier gain and determine the bandwidth.   (Is the bandwidth greater or less than its common emitter counterpart of Chapter 8?)

BW = _____

Activity *BUFFER*

Activity *BUFFER* determines the characteristics of the typical FET buffer of Figure 15.3.

Hand Calculations

9. Using the equations listed below on Figure 15.3, calculate by hand the following. (Because we are still using half-current biasing, g_m at the Q point is the same as determined in step 3.)

A	=	$\dfrac{RL \parallel RS}{RL \parallel RS + 1/gm}$	=	_____
Z_{IN}	=	$RB1 \parallel RB2 \parallel Z_{IN}(FET)$	=	_____
Z_{OUT}	=	$RS \parallel 1/gm$	=	_____

PSpice Analysis

10. Add schematic *BUFFER* to project *fetampbuffer*, and draw the JFET buffer circuit of Figure 15.3.

11. Add the following simulation profiles to schematic *AMPLIFIER*:

   ▪ A *Transient* profile from 0 to 2ms, with a ceiling of 2μs.

   ▪ An *ACsweep* profile from 10kHz to 10GHz, with 100 points/decade.

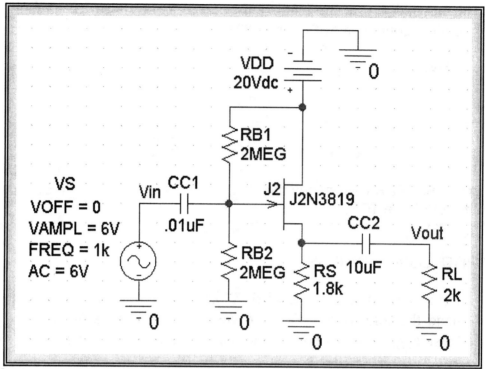

FIGURE 15.3

JFET buffer

12. Using PSpice (transient or AC mode), determine the following. Compare your answer to the theoretical results of step 9.

A (midband)	=	_____
Z_{IN} (midband)	=	_____
Z_{OUT} (midband)	=	_____

13. Using *AC Sweep* mode, determine the buffer's low-frequency cutoff. (The high-frequency cutoff is well beyond 100THz, where $T = 10^{12}$.)

$$f_{CUTOFF} = \underline{\hspace{3cm}}$$

14. Based on the results so far, compare the characteristics of the JFET amplifier/buffer of this chapter with their bipolar counterparts of Chapters 8 and 9. Which one has:

    a. the greatest gain?

    b. the greatest Z_{IN} or Z_{OUT}?

    c. the greatest bandwidth?

Advanced Activities

15. Determine the total harmonic distortion of both the FET amplifier and buffer of this chapter. (*Hint*: See Chapter 19, Volume I.)

16. By performing a parametric analysis on the amplifier of Figure 15.1, vary *RS* over a range of values and determine its effect on circuit gain.

Exercises

1. Analyze the circuit of Figure 15.4, and predict the waveforms at the gate and drain of the FET. (*Hint*: The circuit is called a "chopper.") Generate the waveforms using PSpice and compare them to your predictions.

2. Perform a complete analysis of the three-stage amplifier of Figure 15.5. Why would the design be appropriate for a hand-held battery-powered megaphone? Why is the front-end stage a JFET amplifier, the middle stage a bipolar amplifier, and the final stage a Class B buffer? What is the voltage gain, power gain, efficiency, bandwidth, etc., of the circuit?

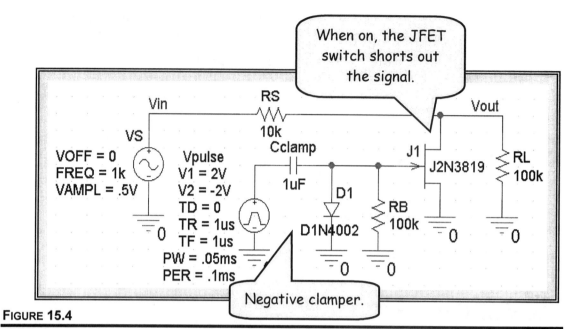

FIGURE 15.4

JFET chopper

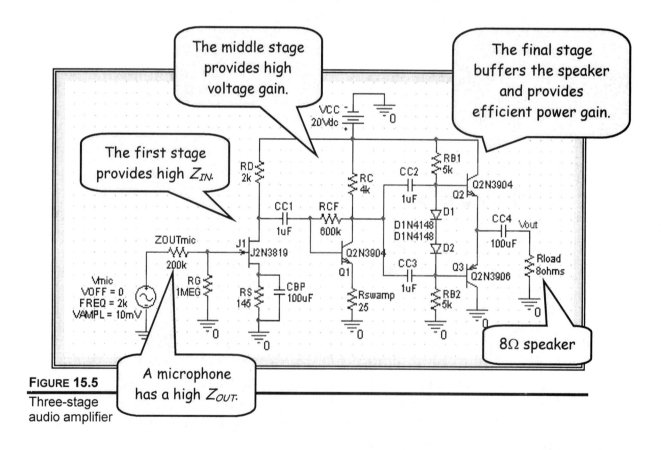

FIGURE 15.5

Three-stage
audio amplifier

Questions and Problems

1.   Why is the voltage gain (A) of an FET amplifier generally less than that of a bipolar amplifier?

2.   Referring to Figures 15.1 and 15.2, why would the voltage gain increase if the value of R_S were decreased?

3.   Referring to Figure 15.1, what allows resistor RB to be so large?   What advantage is there to having a large value of RB?

4.   Why is the power gain of an FET circuit very large?

5.   What is the major difference between bipolar and FET amplifiers and buffers?

Part 4

Special Solid-State Studies

In Part 4 we examine several special applications of solid-state devices. We will find that the transistor can be used as a switch for digital applications and that a special four-layer device can act as a latch.

CHAPTER

16

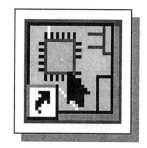

The Transistor Switch

Frequency of Operation

Objectives

- *To design and analyze transistor switches*
- *To increase the frequency of operation*

Discussion

Most of the previous chapters have concentrated on the transistor as used in analog (linear) applications. The other side of the coin is digital (nonlinear) applications. In a digital application, the transistor is a switch that operates between high and low states.

- When the bipolar transistor is used as a switch, the high and low states usually correspond to saturation and cutoff.

- When the JFET is used as a switch, the high and low states usually correspond to I_{DSS} (maximum current) and V_{GSOFF} (no current).

Switching Speed

An important consideration in digital circuits is the time it takes to switch between states. The faster the switching speed, the higher the frequency of operation. For the bipolar transistor, we define the terms of Figure 16.1 as follows:

- t_S (*storage time*) is the time required to come out of saturation (0% to 10%).

- t_R (*rise time)* is the time required to make the transition from saturation to cutoff (10% to 90%).

- t_D (*delay time*) is the time required to come out of cutoff (100% to 90%).

- t_F (*fall time*) is the time required to make the transition from cutoff to saturation (90% to 10%).

For the 3904 transistor, the specification sheet lists these terms as follows:

$$t_S = 200ns \quad t_R = 35ns \quad t_D = 35ns \quad t_F = 50ns$$

The maximum frequency of operation $(1/(t_D + t_R + t_S + t_F))$ is therefore 3.125MHz.

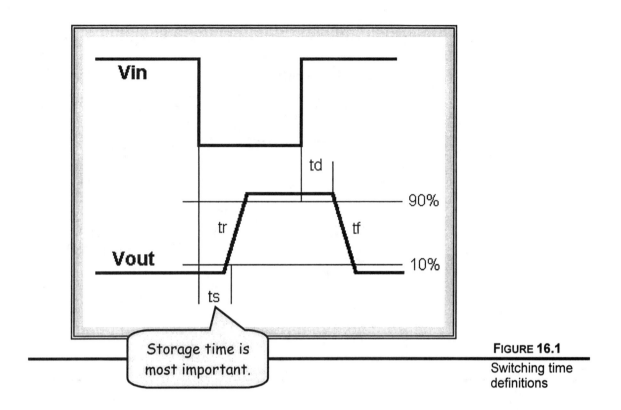

FIGURE 16.1

Switching time definitions

Simulation Practice

Activity *BIPOLAR*

Activity *BIPOLAR* determines the switching speed of a bipolar transistor.

1. Create project *switch* and schematic *BIPOLAR*, and draw the basic transistor switch of Figure 16.2. (*Note*: The values of R_B and R_C were chosen to match the specification sheet test conditions of $I_C MAX = 10$mA and $I_B MAX = 1$mA.)

2. Set the simulation profile, run PSpice, and generate the waveforms of Figure 16.3. (Because of a slight overshoot, you may have to adjust the Y-axis for a 0 to 5V range.)

PSpice for Windows

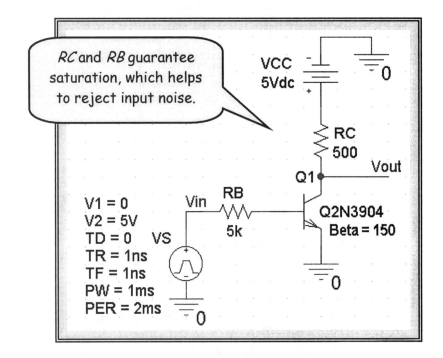

FIGURE **16.2**

Basic bipolar
transistor switch

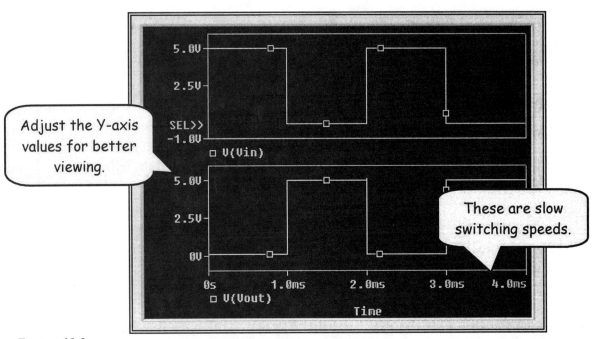

FIGURE **16.3**

Bipolar switch
waveforms

3. Based on the waveforms of Figure 16.3, why is the switch also called an *inverter*?

4. Judging from the waveforms of Figure 16.3, does the transistor operate properly between saturation and cutoff?   (Is the high voltage near 5V and the low voltage near 0V?)

        Yes   No

BJT Switching Time

5. Looking at the waveforms of Figure 16.3, it appears that the output changes instantly with the input. However, increase the frequency by a factor of 1000 (pw = 1µs, per = 2µs), and generate the curves of Figure 16.4.

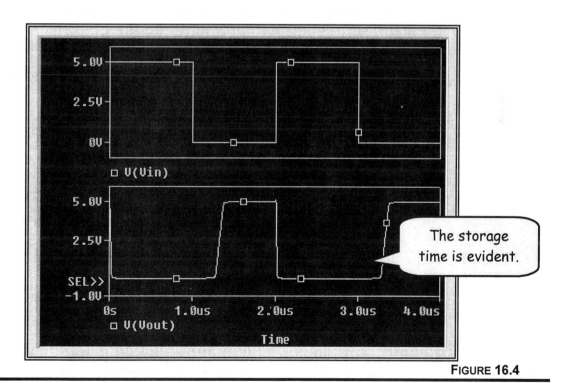

FIGURE 16.4

High-speed waveforms
showing delays

6. Using the cursor, determine values for each of the following and compare them to the specification sheet values listed earlier in the discussion.

$t_S =$ _____ $t_R =$ _____ $t_D =$ _____ $t_F =$ _____

> *Note*: Because our test conditions are not precisely the same as the specification sheet test conditions, do not expect close correlation between the PSpice results and the specification sheet results.

7. To reduce the switching times (especially the storage [t_S] and delay [t_D] times), add the *speedup capacitor* of Figure 16.5. (The speedup capacitor bypasses resistor R_B during *changes*, thereby allowing electrons to move more quickly into and out of the base.)

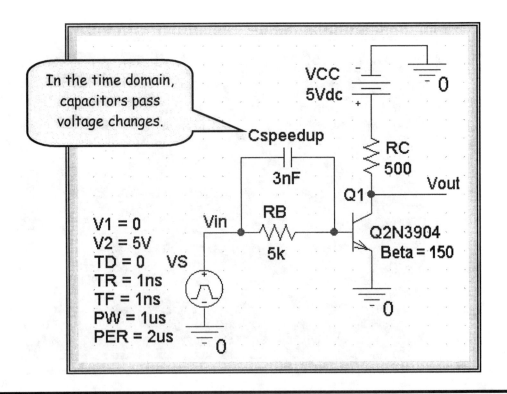

FIGURE 16.5

Adding a speedup capacitor

8. Generate new waveforms and note the storage, delay, and transition times.

    a. Were they greatly reduced?

    Yes   No

    b. However, did spikes appear at the switching points?

    Yes   No

Advanced Activities

9. Draw the JFET switch of Figure 16.6, determine an appropriate value of R_D, and generate input/output waveforms. (*Hint*: Refer to Figure 12.3 and note that the transistor is driven between the top and bottom curves as V_{GS} alternates between 0 and –3V.)

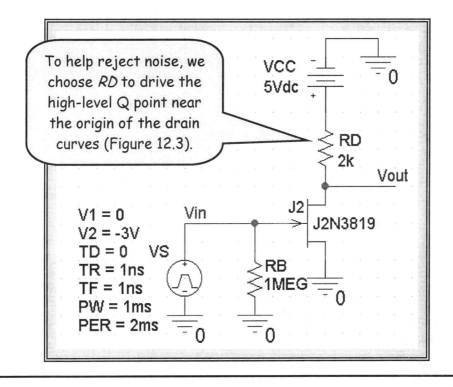

FIGURE **16.6**

JFET switch

10. Referring to step 8, find a way to reduce the "speedup" spikes. (*Hint*: Consider the use of regular and zener diode limiter circuits.)

11. Using the equation developed in the discussion, analyze the "speedup" waveforms generated by step 8 (expand the rise and fall sections) and determine the highest frequency of operation. By increasing the frequency in steps, verify your findings using PSpice.

12. Compare the base current between the regular and speedup switches (Figures 16.2 and 16.5) by generating weaveforms. Comment on the results.

Exercises

1. Use PSpice to determine the maximum frequency of operation of the CMOS inverter of Figure 16.7 and compare to step 11. (Could this basic inverter be part of a 500MEGHz Pentium microprocessor?)

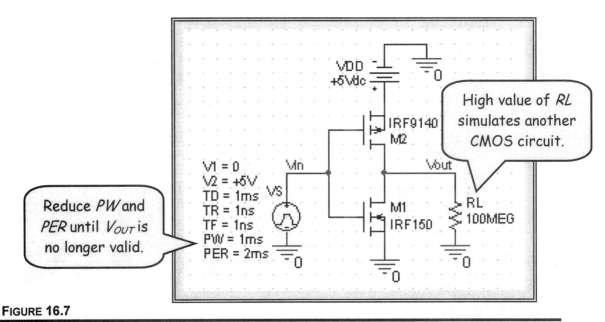

FIGURE 16.7

CMOS inverter

2. For the CMOS inverter of the previous exercise, determine how the power and energy requirements change with frequency. What does this say about the heat sink requirements of high-frequency switches?

3. Determine the input/output characteristics and maximum frequency of operation of the TTL (*transistor-transistor logic*) inverter of Figure 16.8. Place speedup capacitors around *R1*, *R2*, and *R4*, and again determine the maximum frequency of operation.

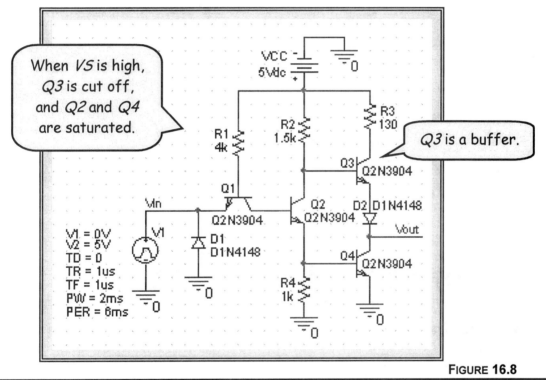

FIGURE 16.8

TTL inverter circuit

4. The ECL (*emitter-coupled logic*) switch (inverter) shown in Figure 16.9 increases the switching speed by avoiding saturation. Explain how the circuit works. Use PSpice to determine its characteristics and maximum frequency of operation. (Does it avoid saturation?)

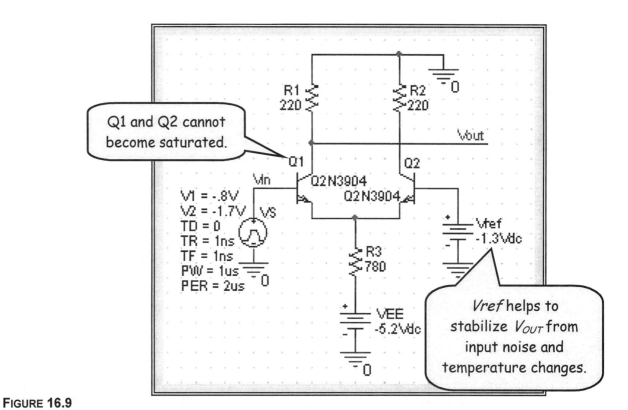

FIGURE 16.9

ECL inverter
circuit

Questions and Problems

1. When the transistor of Figure 16.2 enters saturation, what happens to the
    following?   (Enter "up" or "down" after each term.)

    a. V_{CE} goes _____
    b. *Beta* goes _____
    c. I_C goes _____

2. Electrons that have saturated the base are swept out of the base during:

    t_S t_R t_D t_F

3.   Why does the speedup capacitor of Figure 16.5 reduce the delay (t_D) and storage (t_S) times?

4.   On the load line graph in Figure 16.10, circle the area in which analog (linear) circuits normally operate, and box the areas in which digital circuits normally operate (except when in transition).

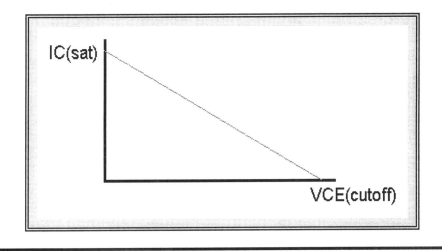

FIGURE 16.10

Analog/digital operating region comparison

5.   Referring to Figure 16.2, how does driving the transistor into saturation help to reject input noise?   (*Hint*: When *VS* = 4V, is the transistor still in saturation?)

6. Prove that neither transistor of Figure 16.9 can ever be in saturation. (*Hint*: What is the worst-case state when V_{CE} of Q1 is minimum?)

17

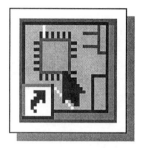

Thyristors

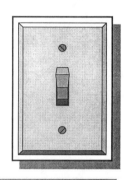

The Silicon-Control Rectifier

Objectives

- *To determine the on and off characteristics of the SCR*
- *To use the SCR in the design of an efficient power controller*

Discussion

Thyristors are used for special switching applications. The most common thyristor is the *silicon-control rectifier* (SCR), a device that uses positive feedback to act as a latch. When turned on, it tends to stay on; when turned off, it tends to stay off.

Turning to Figure 17.1(a), an SCR is a four-layer PNPN device. As shown by the equivalent circuit of Figure 17.1(b), it acts as overlapping NPN and PNP transistors. Because the collector of one transistor feeds the base of the other, both transistors must be on or both must be off. (It is impossible for one transistor to be on and the other off, except during the brief time when they are changing state.)

The schematic symbol for the SCR is given in Figure 17.1(c). (Because the SCR is made up of PSpice primitives, it is a *subcircuit* and therefore carries the "X" designator.)

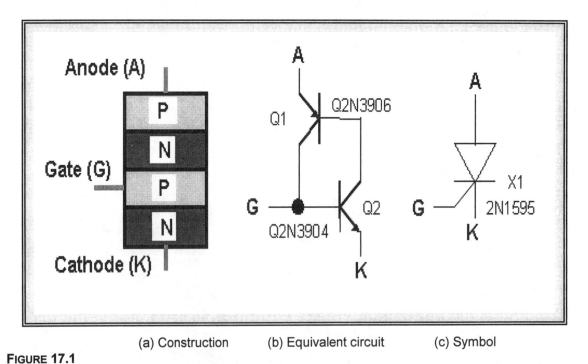

(a) Construction   (b) Equivalent circuit   (c) Symbol

FIGURE 17.1

The SCR

We will use the test circuit of Figure 17.2 to demonstrate the SCR's characteristics.

- The SCR is turned on by a threshold combination of anode (*A*) voltage and gate (*G*) current.

- The SCR is turned off when the forward anode-to-cathode current drops below the *holding current* (I_H).

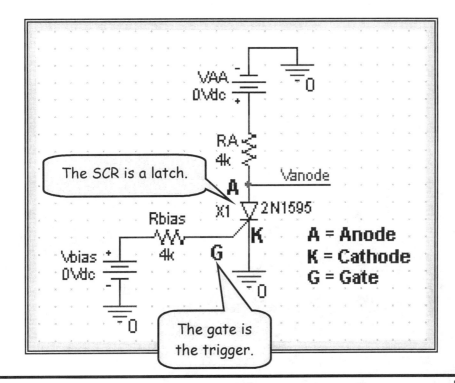

FIGURE 17.2
SCR test circuit

Simulation Practice

Activity *SCR*

Activity *SCR* will use the test circuit of Figure 17.2 to generate SCR operating curves.

1. Create project *thyristors* and schematic *SCR*, and draw the test circuit of Figure 17.2. (Because both V_{BIAS} and V_{AA} will be swept, DC bias point values need not be assigned at this time.)

SCR Operating Curves

2. Create simulation profile DCsweep as follows:

   - The primary sweep variable is V_{AA} from 0 to 55V in increments of 1V.

   - The secondary sweep variable is V_{BIAS}, with values of 8.35, 8.5, and 8.75.

3. Generate the SCR *operating curves* of Figure 17.3. (Be sure to change the X-axis variable to *V(Vanode)*).

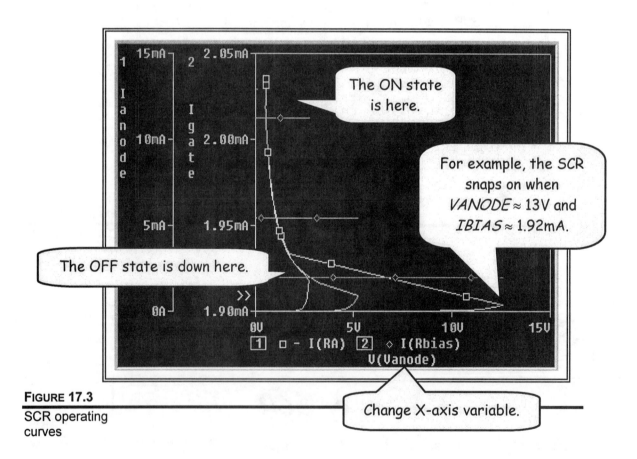

FIGURE 17.3
SCR operating curves

4. List below the three pairs of anode voltages (*V(Vanode)*) and gate currents (*I(Rbias)*) shown in Figure 17.3 that will turn on the SCR.

	$V(V_{ANODE})$	$I(R_{BIAS})$
Pair 1		
Pair 2		
Pair 3		

Activity *POWERCONTROLLER*

Activity *POWERCONTROLLER* shows how the SCR can efficiently control the average power to a load.

5. Create schematic *POWERCONTROLLER*, and draw the circuit of Figure 17.4. (Resistor *Rcontrol* varies the percentage of time that the SCR is on by controlling the trigger current.)

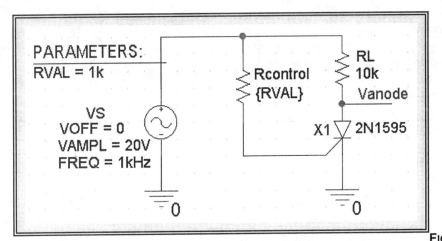

FIGURE 17.4

SCR power control

6. Set the simulation profile for a primary transient sweep to 2ms and a parametric sweep of *Rcontrol* by the values 1k, 5k, and 10k.

7. Run PSpice and generate the curves of Figure 17.5. (Note the use of the "@" operator and numbers to single out individual parametric curves.)

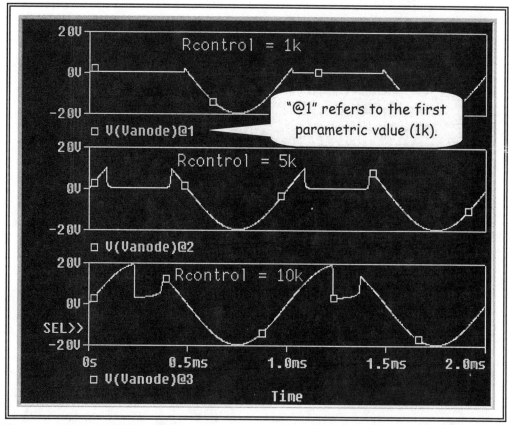

FIGURE 17.5

SCR power control
waveforms

8. Based on Figure 17.5, what value of *Rcontrol* delivers the greatest power to the load? (*Hint*: Power is delivered to the load when the SCR is on.)

1kΩ　　　　5kΩ　　　　10kΩ

Advanced Activities

9. For the *Rcontrol* = 10kΩ case, determine the approximate holding current. (*Hint*: What is the current when the SCR *starts* to turn off?)

10. Design an SCR power controller for a square wave. (*Hint*: Make use of a capacitor to vary the turn-on delay time.)

Exercises

1. Substitute the SCR equivalent circuit of Figure 17.1(b), and re-generate the curves of Figure 17.5. Is the equivalent circuit similar in function to the SCR?

2. A *triac* is a latching device that consists of two back-to-back SCRs to achieve full-wave control. Using the 2N5444 triac, redesign the power control circuit of Figure 17.4 and compare your results with those of Figure 17.5.

Questions and Problems

1. How do you switch an SCR on, and how do you switch it off?

2. Referring to Figure 17.1(b):

    a. Explain how positive feedback comes into play during switching.

  b. Why are both of the transistors either on or off?

3. Based on Figure 17.3, if the gate current is 1.93mA, approximately what anode voltage would turn on the SCR?

4. Why is the SCR an efficient method of controlling power? (*Hint*: Is any power wasted in a resistor?)

5. What is the difference between the SCR and the triac in the two symbols below?

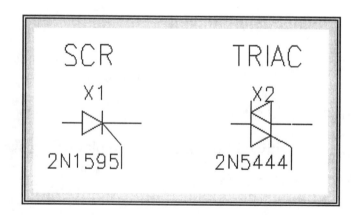

6. Referring to Figure 17.4, when the SCR is on, how much power is delivered to the load?

Part 5

Operational Amplifiers

In Part 5 we concentrate on the science of cybernetics, where the power of feedback gives conventional circuits amazing properties. We will find that the differential amplifier is the ideal circuit to implement feedback.

A precision differential amplifier is known as an operational amplifier and is one of the most powerful analog devices available to the designer. When the operational amplifier is combined with negative and positive feedback, we create such devices as *amplifiers*, *buffers*, *integrators*, *differentiators*, *oscillators*, and *filters*.

18

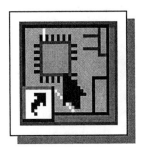

The Differential Amplifier

Common-Mode Rejection

Objectives

- *To analyze the differential amplifier*
- *To compare the differential mode with the common mode*

Discussion

A differential amplifier has two inputs. It strongly amplifies the *difference* between the two inputs, but it rejects signals that are *common* to both inputs. Such an amplifier is shown in Figure 18.1.

- Signals common to both inputs are forced to pass through the large 5kΩ emitter bias resistor and are rejected.

- Signals differential to both inputs pass easily through two small *re'* values and are amplified.

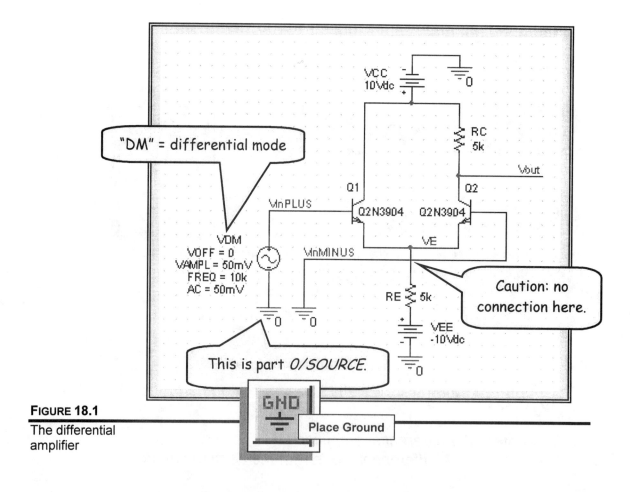

FIGURE 18.1

The differential amplifier

As an added bonus, the differential amplifier of Figure 18.1 requires no coupling or bypass capacitors and is, therefore, a *DC amplifier* (one with no low-frequency rolloff).

The split power supply eliminates the input coupling capacitor by placing the input leads at ground, and the symmetric arrangement of the transistor pair eliminates the bypass capacitor because each transistor bypasses the other.

As we will see in the following chapters, the differential amplifier provides a far more important function than simply amplifying differential-mode DC signals: it provides a convenient method of implementing *negative or positive feedback.*

Because the operational amplifiers of Part 5 may be covered in a separate course, we have included in this chapter reviews of the most fundamental PSpice techniques. (Also see Appendix A.)

Simulation Practice

Activity *AMPLIFIER*

Activity *AMPLIFIER* analyzes the typical, discrete differential amplifier of Figure 18.1.

Hand Calculations

1. For the amplifier of Figure 18.1, calculate each of the following. (Assume $\beta \approx 175$ and $re' \approx 25\text{mV}/I_{EQ}$.)

$I_{CQ} =$	
$V_{CEQ} =$	
$A_{DM} =$	
$Z_{IN} =$	
$Z_{OUT} =$	

PSpice Analysis

2. Create project *differential*.

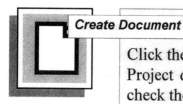

	Review

Click the *Create document* toolbar button to bring up the New Project dialog box. Enter *differential* in the Name box, and check the Location (Is the pathway correct?).

OK to bring up the Create PSpice Project dialog box. **Create a blank project**, **OK**, to bring up the Project Manager and Edit windows.

3. Rename the schematic name from the default *SCHEMATIC1* to *AMPLIFIER*.

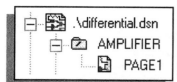

Review

CLICKR on *SCHEMATIC1*, **Rename**, enter *AMPLIFIER*, click anywhere to set the change.

4. Draw the circuit of Figure 18.1.

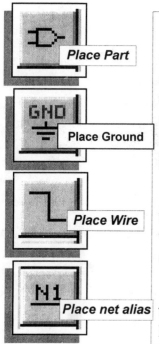

Review

CLICKL on the *Place Part* toolbar button (**Add Library** if necessary), select the desired *library*, select the desired *part*, **OK**, drag to desired location, **CLICKL** to place (again if second instance of part is desired), **CLICKR** to end mode. To add grounds, click the *Place Ground* toolbar button (**Add Library** *source* if necessary).

To wire the parts together, **CLICKL** on the *Place Wire* toolbar button, **CLICKL** the beginning wire location, drag wire, **CLICKL** the ending location (again for other wire connections), **CLICKR** to end mode.

To name a wire (net), **CLICKL** on the *Place net alias* toolbar button, enter the desired name, drag to desired location, **CLICKL** to place, **CLICKR** to end mode.

5.   Set the simulation profile to *Bias Point*.

> **Review**
>
> **CLICKL** on the *New Simulation Profile* toolbar button, enter name *BiasPoint*, **Create**, select Analysis Type *Bias Point*, **OK**.

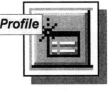

6.   Run PSpice and enter the transistor Q point DC values below.

I_{CQ} (I(V$_{CC}$)/2) _____ V_{CEQ}(Vout–V$_E$) = _____

Do the PSpice values approximately match those calculated in Step 1?

   Yes   No

7.   Switch to a *transient* mode (0 to .2ms), with increments of .2μs.

> **Review**
>
> **CLICKL** on the *New Simulation Profile* toolbar button, enter name *Transient*, **Create**, select *Analysis Type* Time Domain, *Run to Time* .2ms, *Start saving data after* 0, *Maximum step size* .2μs, **OK**.

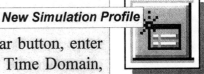

8.   Using PSpice, determine midband A$_{DM}$ (differential mode gain), Z$_{IN}$, and Z$_{OUT}$, and place your answers on the following page.

> **Review**
>
> To determine Z_{OUT}, add a capacitor to the output, and use either of the following methods:
>
> a.   **Algebra**: Measure V_{OUT} with and without a load. Use algebra to determine Z_{OUT}.
>
> b.   **Injection**: Ground the inputs, move *VSIN* to the output, and measure V_{OUT} and I_{OUT} directly.

$$A_{DM} = \underline{\hspace{4cm}}$$

$$Z_{IN} = \underline{\hspace{3.5cm}}$$

> Use peak-to-peak/2 where appropriate to factor out the DC offset.

$$Z_{OUT} = \underline{\hspace{3.5cm}}$$

Do the PSpice values approximately match those calculated in step 1?

Yes   No

9. Based on the phase relationship between the input and output voltages, circle the appropriate label for V_{IN1}.

Inverting   Noninverting

10. Move V_{DM} from V_{IN1} to V_{IN2}, again measure the phase relationship between input and output, and circle the appropriate label for V_{IN2}.

Inverting   Noninverting

11. Using the symbols for *inverting* (−) and *noninverting* (+), label the two inputs of Figure 18.1.

12. Set an *AC Sweep* simulation profile from 1Hz to 100MEGHz (100 points/decade) and determine the bandwidth of the amplifier.

$$BW = \underline{\hspace{4cm}}$$

Is it a DC amplifier?   (A DC amplifier has no low-frequency rolloff.)

Yes   No

Common Mode

13. Connect the circuit in the *common mode*, as shown in Figure 18.2.

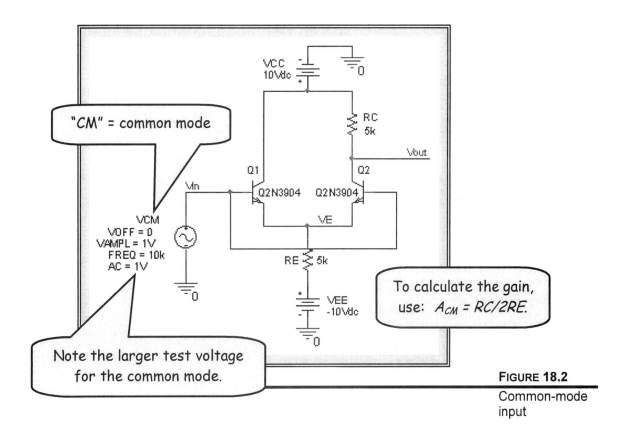

FIGURE 18.2

Common-mode
input

14. Using the test mode of your choice (*Transient* or *AC Sweep*), measure the midband common-mode voltage gain and enter the result below.

$$A_{CM} = \underline{\hspace{3cm}}$$

Comparing A_{DM} to A_{CM}, does the differential amplifier tend to reject common-mode input signals?

Yes   No

15. As a measure of how much better differential-mode signals are amplified over common-mode signals, determine the *common-mode rejection ratio* (CMRR).

$$CMRR = \frac{A_{DM}}{A_{CM}} = \underline{\hspace{3cm}}$$

Advanced Activities

16. Based on the values of A_{DM} and A_{CM} determined earlier, sketch a time-domain graph of V_{OUT} for the circuit of Figure 18.3. Compare your answer with the measured (PSpice) results.

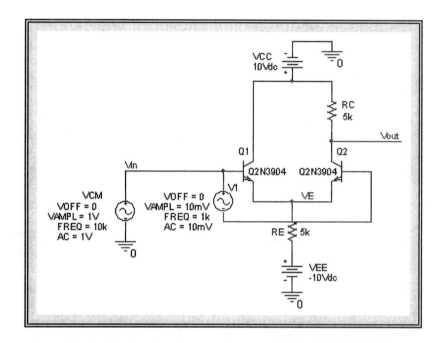

FIGURE 18.3

Test circuit

17. The circuit of Figure 18.4 uses *current-source* biasing to improve CMRR. Measure the circuit's CMRR and compare your results to those of step 15. Did the CMRR go up?

Yes   No

18. Measure the harmonic distortion of any of the amplifiers of this chapter. Would you say it is high or low? (Assume that any value above 10% is high.)

High   Low

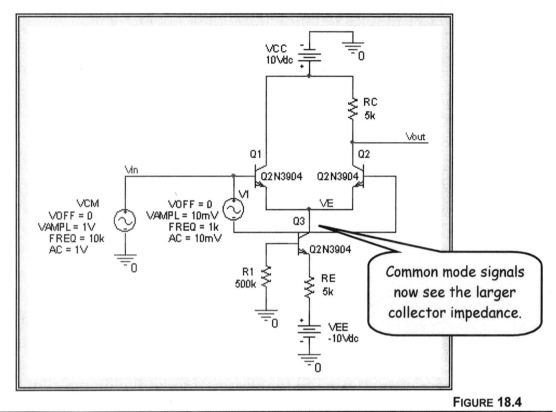

FIGURE 18.4

Using current-source biasing

Exercises

1. Design a differential amplifier that gives a differential (balanced) output. (Hint: Add a 5k resistor to the collector of Q1 in Figure 18.4, and tap the voltage off both *Q1* and *Q2*.) How does the gain compare to that of Figure 18.4?

2. Design a differential amplifier with a CMRR of approximately 100.

3. Run a Monte Carlo analysis (Chapter 28) on resistors *RE* and *RC* of Figure 18.1 separately and compare the results.

4. Compare the harmonic distortion between the amplifiers of Figures 18.1 and 18.4.

PSpice for Windows

Questions and Problems

1.   Why is a *bypass* capacitor not required in the differential amplifier?

2.   Why is there a "2" in each of the gain equations below?

Differential-mode gain   Common-mode gain

$$A_{DM} = RC/2re'$$ $$A_{CM} \cong RC/2RE$$

3.   How could you swamp the circuit of Figure 18.1 in order to reduce the harmonic distortion?   What would happen to the voltage gain (A_{DM})?

4.   When the circuit was modified as in Figure 18.4, why did the CMRR go up?

5. Given the following, determine V_{OUT}*(peak)* as accurately as you can.   (*Note*: Assume that V_{IN1} and V_{IN2} are peak values and are the same frequency and phase.)

$V_{IN1} = 200mV$, $V_{IN2} = 215mV$, $A_{DM} = 100$, $A_{CM} = 1$

$V_{OUT} = $ _____

19

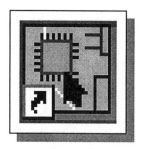

Negative Feedback

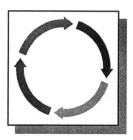

Open Loop Versus Closed Loop

Objectives

- *To design a three-stage amplifier*
- *To compare the properties of an open-loop and closed-loop amplifier*

Discussion

In this chapter we show how a magical drop of negative feedback can greatly enhance the properties of an amplifier. We start with the conventional three-stage amplifier of Figure 19.1—an amplifier that presently does not employ feedback (the *open loop* configuration).

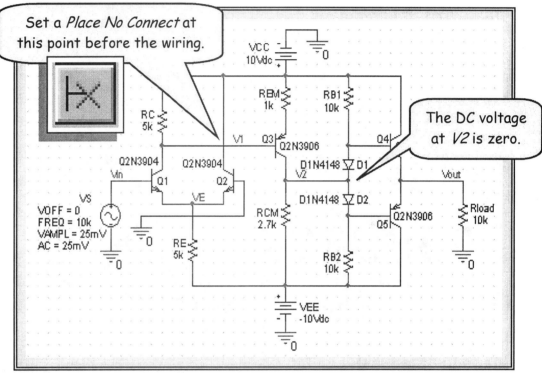

FIGURE 19.1

Three-stage amplifier
(open loop)

By looking more closely at the open-loop configuration of Figure 19.1, we notice a differential input stage, a common-emitter middle stage, and a Class B buffer final stage. We choose the differential amplifier of Chapter 18 for the first stage, because it is ideally suited to carry out negative feedback. The Class B buffer adds a good measure of power gain to the circuit.

We note the complete absence of coupling and bypass capacitors, making the amplifier a DC amplifier (able to amplify zero hertz input signals).

- Coupling capacitors are not required because of the split power supply and because the middle stage is also a *level-shifter* that allows for *direct coupling* between stages.

- Bypass capacitors are not required because the first stage is a differential amplifier, the second stage is a level-shifter, and the third stage is a buffer.

Closing the Loop

Then, for just 5 cents in added components (two resistors), we add negative feedback and close the loop (Figure 19.2). What results is nothing short of miraculous: an amazing improvement in design ease, stability, distortion, bandwidth, Z_{IN} and Z_{OUT}.

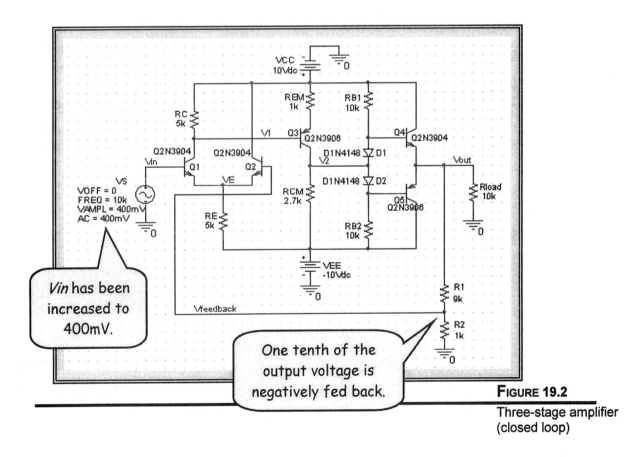

FIGURE 19.2

Three-stage amplifier (closed loop)

Simulation Practice

Activity *AMPLIFIER*

Activity *AMPLIFIER* compares the properties of an open loop (Figure 19.1) versus closed-loop (Figure 19.2) amplifiers.

1. Create project *negativefeedback* with schematic *AMPLIFIER*.

2. Draw the three-stage open-loop (OL) amplifier of Figure 19.1, and set the attributes as shown. (We tap the first-stage signal off Q1 to compensate for the phase inversion introduced by Q3.)

3. Create both of the following profiles:

    -   *Transient* from 0 to .2ms in increments of .2μs. Enable the Fourier option for *Vout*.

    -   *ACsweep* from 10Hz to 1G, with 100 pts/decade. Enable the noise option, with Output Voltage of *V(Vout)*, an I/V Source of *VS*, and an interval of *1kHz*.

4. Using PSpice (transient, Fourier, and AC modes as needed), determine the following open-loop (OL) values. (Consider HD and SNR to be optional.)

A (gain)	=	_____ A_{dB} = _____
Z_{IN}	=	_____
Z_{OUT}	=	_____
BW (bandwidth)	=	_____
HD (distortion)	=	_____
SNR (signal/noise)	=	_____

Optional

Closed Loop (With Feedback)

5. Add the negative feedback path shown in Figure 19.2.

6. First, use the *transient* mode to look at the input and output waveforms. Is the output "well behaved" (no clipping or distortion), and is the voltage gain approximately 10?

    Yes   No

7. Switch to the *AC Sweep* mode and generate the gain and phase plots of Figure 19.3.

    > The dramatic spike at approximately 21MHz results from (1) negative feedback turning into positive feedback due to high-frequency phase shifts and (2) overall gain > 1.

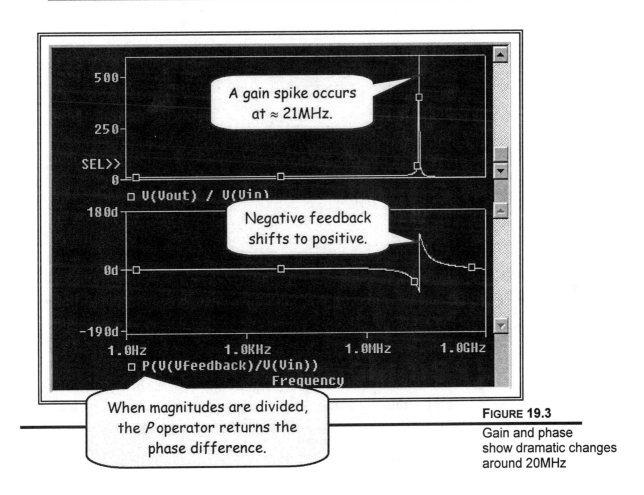

FIGURE 19.3

Gain and phase show dramatic changes around 20MHz

PSpice for Windows

8.   By how much does the phase shift at the critical point?

     Phase change = _____

9.   To remove the high-frequency gain spike, add the .5nF *compensation capacitor* (CC) shown in Figure 19.4.

> The compensation capacitor creates a low-pass filter that reduces the overall gain below 1 when the feedback turns positive at high frequencies. This prevents the output signal from growing and producing a spike.

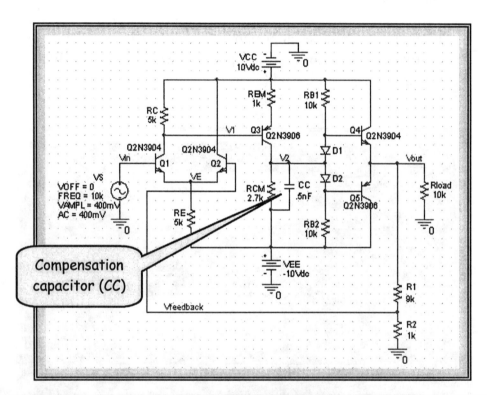

FIGURE 19.4

Compensation
capacitor added

10. We again generate plots of gain and phase shift (Figure 19.5). Have the high-frequency gain spike and sudden phase shift been removed, and is the midband gain approximately 10?

           Yes   No

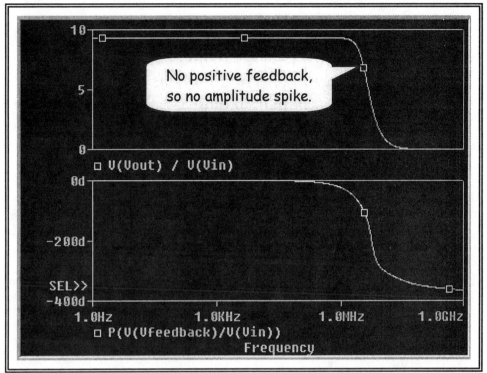

FIGURE 19.5

Waveforms with compensation

11. Use PSpice to determine each of the following closed-loop (CL) values on our newly designed, compensated amplifier. (As before, all AC values are midband.)

A (gain)	=	_____ A_{dB} = _____
Z_{IN}	=	_____
Z_{OUT}	=	_____
BW (bandwidth)	=	_____
HD (distortion)	=	_____
SNR (signal/noise)	=	_____

Optional

PSpice for Windows

12. Does the phase angle between *Vin* and *Vfeedback* go from 0° to 360° after the breakpoint?

Yes   No

13. As a final summary and comparison, enter the specifications of each type of circuit side by side below. (For a fair comparison, go back to Figure 19.1, add the .5nF compensation capacitor across RCM, and remeasure the open-loop bandwidth.)

		Open loop (OL)	Closed loop (CL)
A(reg)	=		
Z_{IN}	=		
Z_{OUT}	=		
BW	=		
HD	=		
SNR	=		

14. Based on the results of step 13, what conclusions can you draw about the effects of negative feedback?   (What do we profit by sacrificing voltage gain?)

Advanced Activities

15. Reduce the load (*Rload*) of both amplifiers (with and without feedback) from 10k to 100Ω and explain the results.

16. (To perform this exercise, review Chapter 29.) Set 10% device tolerances on all the resistors of the feedback amplifier of Figure 19.4, except the two feedback resistors (*R1* and *R2*). Run a worst-case analysis and note the results. Then reverse the configuration (only *R1* and *R2* have 10% tolerances) and note the results. By comparing the two cases, what conclusions can you draw? (To improve sensitivity, would it be a waste of money to upgrade *all* resistors to 1% tolerance?)

17. Change the value of the compensation capacitor (CC) and explain the results.

Exercises

1. Design a negative feedback amplifier that uses 100% feedback. What are its major characteristics? (Would it make a good buffer?)

2. Change the feedback ratio (presently 9 to 1) over several values and re-measure the gain each time. Can you create a formula for closed-loop gain?

3. Remove (or bypass) the class B buffer output stage from the circuit and explain the results.

Questions and Problems

1. To achieve negative feedback, the feedback signal must be:

   a. in phase with the input signal.
   b. $180°$ out of phase with the input signal.

2. What is a *direct-coupled* amplifier?

3.   Why is the final stage of the amplifier of Figure 19.1 a Class B buffer?

4.   How is the gain of the closed-loop amplifier of Figure 19.4 related to the values of the feedback resistors (*R1* and *R2*)?   How is it related to the percentage of feedback?   (See Exercise 2.)

5.   Why is the closed-loop amplifier of Figure 19.4 a better *buffer* than the open-loop amplifier of Figure 19.1?

6.   Did the addition of a compensation capacitor to the closed-loop configuration (Figure 19.4) cost us any significant bandwidth?

7.   What is the difference between *negative* and *positive* feedback?

20

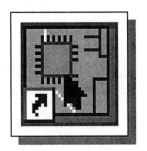

The Operational Amplifier

Specifications

Objective

- *To compare test results of the 741 op amp with its major specifications*

Discussion

The homemade differential amplifier of the last chapter has a number of drawbacks. First, the gain and input impedance are too low, and the output impedance is too high. There are also many other secondary characteristics we wish to improve.

In this chapter, we will examine a commercially available operational amplifier and see if it offers the improved characteristics we desire. Our choice is the 741 workhorse op amp, supplied within PSpice library *eval.slb*.

Specifications

The characteristics of any circuit, such as the 741 op amp, are known as *specifications*. The important specifications of the 741 op amp are listed in Table 20.1. Using PSpice we will verify each of these one at a time. (If time is short, verify those of primary interest first.)

741 Specifications	
• Input offset voltage	1mV
• Maximum/minimum output voltages	14V (15V supplies)
• Open loop voltage gain	200,000 (106dB)
• Input bias current	80nA
• Input offset current	20nA
• Short circuit output current	25mA
• Input impedance	2MEG
• Output impedance	75ohms
• CMRR	90DB (30,000)
• Slew rate	.5V/μs
• Break frequency	10Hz
• Frequency roll off	20dB/Dec
• Gain-BW product	1MEGHz

TABLE 20.1

741 op amp
specifications

Of all the specifications listed, perhaps the most surprising is the very low value of the break frequency (10Hz). However, as we learned in Chapter 19, this is done deliberately in order to force the overall closed-loop op amp gain below one before phase shifts cause oscillations by way of positive feedback.

Simulation Practice

Activity *OPAMP*

Activity *OPAMP* uses the circuit of Figure 20.1 to test the specifications of the 741 operational amplifier.

1. Create project *specification* with schematic *OPAMP*.

2. Draw the test circuit of Figure 20.1.

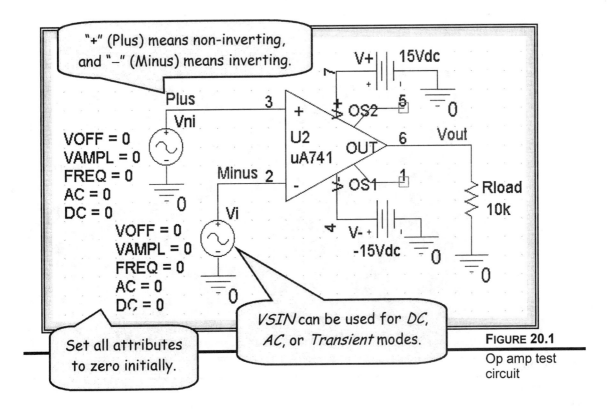

FIGURE 20.1
Op amp test circuit

3.   Set up the following simulation profiles:

▪ A *DC Sweep* of *Vni* from –200uV to +200uV, in steps of .1µV.

▪ A *transient* analysis (0 to .2ms), with a step ceiling of .2µs.

▪ An *AC Sweep* of *Vni* from 1Hz to 1GHz, with 100 points/decade.

Input offset voltage — *The voltage at either input that drives V_{OUT} to zero.*

4.   Perform a DC Sweep and display the V_{OUT} curve of Figure 20.2. Determine V_{OFF}, the value of Vni that drives V_{OUT} to zero. Compare your result with the specification sheet value from Table 20.1.

V_{OFF} (PSpice) = _____ V_{OFF} (spec) = ___1 mV___

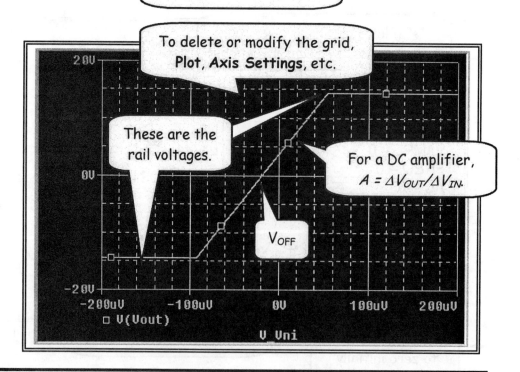

FIGURE 20.2

DC Sweep
curves

Maximum/minimum V_{OUT} values — *Also known as the rail voltages, the maximum and minimum output voltages allowed for a particular power supply level (15V in our case).*

5. Using the graphical results of Figure 20.2, enter the rail voltages.

$V_{RAIL}+$ (PSpice) = _____ $V_{RAIL}+$(spec) = __+14V__

$V_{RAIL}-$ (PSpice) = _____ $V_{RAIL}-$(spec) = __–14V__

Open-loop voltage gain — *The DC differential voltage gain without feedback.*

6. By measuring the slope of the input/output curve of Figure 20.2, record below the open-loop voltage gain ($\Delta V_{OUT}/\Delta V_{IN}$).

A_{OL}(PSpice) = _____

A_{OL}(spec) = ___200,000 (106dB)___

7. Based on the DC Sweep mode:

a. Is this gain the DC or low-frequency gain?

Yes   No

b. Is this gain value necessarily valid at high frequencies?

Yes   No

Short circuit output current — *The output current when V_{OUT} is shorted.*

8. Returning to Figure 20.1, short the output to ground (set RL to a very small value) and plot the output current ($I(Rload)$) using the DC Sweep methods of step 2. The short circuit current is the absolute maximum current. When done, return *Rload* to 10k.

I_{SHORT} (PSpice) = _____ I_{SHORT} (spec) = __25mA__

Input bias current — *The average value of the input currents (Vni and Vi grounded).*

9. Using *bias point* techniques (bias currents), determine *I(Vni)* and *I(Vi)*, and use the equation below to calculate the input bias current. (Make sure that DC = 0 for both inputs.)

$$I_{BIAS} \text{ (PSpice)} = \frac{I_{NI} + I_I}{2} = \underline{\hspace{2cm}} \qquad I_{BIAS} \text{ (spec)} = \underline{80nA}$$

Input offset current — *The difference between the input bias currents.*

10. Using the data from Step 9, determine the input offset current.

$$I_{OFFSET}(\text{PSpice}) = I_{NI} - I_I = \underline{\hspace{2.5cm}} \qquad I_{OFFSET}(\text{spec}) = \underline{20nA}$$

Input impedance — *The total resistance between the inputs.*

11. Set *Vni*'s VAMPL to 1mV and FREQ to 10k (leave Vi grounded). Using transient analysis, measure *I(Vni)* and determine Z_{IN} using Ohm's law. (*Hint*: Use peak-to-peak to cancel out the DC bias current offset.)

$$Z_{IN} \text{ (PSpice)} = \underline{\hspace{2.5cm}} \qquad Z_{IN} \text{ (spec)} = \underline{2MEG}$$

Output impedance — *The resistance viewed from the output terminal.*

12. Using the same transient analysis as step 11, plot V_{OUT} with and without a load (i.e., $RL = 100\Omega$ and $100MEG\Omega$.) Use algebra to determine Z_{OUT}.

$$Z_{OUT} \text{ (PSpice)} = \underline{\hspace{2.5cm}} \qquad Z_{OUT} \text{ (spec)} = \underline{75\Omega}$$

Common-mode rejection ratio (CMRR) — *Open-loop differential gain (A_{DM}) divided by common mode gain (A_{CM}).*

13. Set up the input for common mode (short both inputs to *Vni*). Perform a DC Sweep between +5V and –5V, display V_{OUT}, and determine A_{CM}. (*Hint*: A_{CM} equals the slope of the output curve.)

A_{CM} = _____

14. Use the data gathered in steps 6 and 13 to determine CMRR.

CMRR(PSpice) = _____   CMRR(spec) = <u>90dB (30,000)</u>

Slew rate — *The slope of V_{OUT} to a step input signal.*

15. Substitute *VPULSE* for *VSIN* (*VNI*) and set up the input for differential mode (Figure 20.1). Program *VPULSE* for a fast rise-time step input of 0 to .1V, and run a transient analysis from 0 to 50µs. (1pF would be a fast risetime.)

16. Display V_{OUT} (to 50µsec) and determine its slew rate (slope) in volts/µsec.

Slew rate(PSpice) = _____   Slew rate(spec) = <u>.5V/µs</u>

17. Do you think the slew rate factor might produce output signal distortions at high frequencies?

Yes   No

Frequency response — *The gain versus the frequency (Bode plot).*

18. Set *Vni* to AC = 50µV (either *VSIN* or *VPULSE* is okay) and leave *Vi* grounded. Perform a logarithmic AC Sweep from 1Hz to 10MEGHz, and generate the graph of Figure 20.3. Answer the following:

a. What is the break frequency?

F_B (PSpice) = _____ F_B (spec) = ___<u>10Hz</u>___

b. What is the rolloff? (*Hint*: Measure *ΔVdB* between 1kHz
    and 10kHz.)

Rolloff(PSpice) = _____   Rolloff(spec) = __20dB/Dec__

c. What is the *gain-bandwidth product*?   (*Hint*: Measure the
    unity gain frequency, the frequency at which the gain is 1,
    or 0dB).

Gain-BW(PSpice) = _____   Gain-BW(spec) = __1MEGHz__

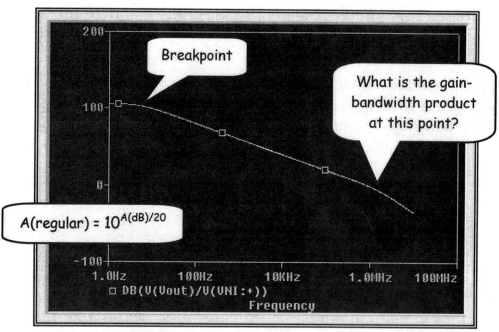

FIGURE 20.3

Open-loop
Bode plot

19. Pick several points along the curve (well past the breakpoint) and
     determine $A \times f$ (gain times frequency). What generalization can
     you make about this value (the same *gain-bandwidth product*
     defined in step 18c)?

Advanced Activities

20.   Repeat any or all of the specification steps using the LM324 op amp.   Compare and summarize your results.

21.   Following up on steps 15-17, assume we have a 10V output signal from a 741 op amp.   Theoretically, what is the highest frequency possible without slew rate distortion?

   Hint: As shown below, we obtain the slope of a sine wave by differentiation.   The frequency at which the maximum slope equals the slew rate is the highest frequency possible without slew rate distortion.   [*Hint*: The maximum value of cos( ) = 1.]

$$.5V/\mu s = \frac{d(10V \sin(2\pi ft))}{dt} = 10V \times 2\pi f \cos(2\pi ft) = 10V \times 2\pi f \,(\text{max})$$

   Solving for f gives = ____7958Hz____

   f is inversely proportional to signal amplitude.

   Using PSpice, determine the harmonic distortion above and below this value and note the waveform shape.   Summarize your results.   (Be sure to adjust the input to always generate a 10V output.)

22.   For the amplifier of Figure 20.1, determine the signal-to-noise ratio at a midband (very low) frequency.

Exercises

1.   Compare the slew rate of the 741, 324, and 711 op amps and determine which one would be most appropriate for digital applications.

2.   Design a *comparator* circuit in which $V_{OUT} = +15V$ for all V_{IN} greater than +5V and $V_{OUT} = -15V$ for all V_{IN} +5V or less.

Questions and Problems

1. Is the circuit of Figure 20.1 in the *open-loop* or *closed-loop* configuration?

2. For many applications why is it possible to ignore many of the specifications of this experiment (such as *input offset current*)?

3. How do the specifications for A, Z_{IN}, and Z_{OUT} for the 741 differ from the characteristics of the discrete amplifier of the previous chapter?

4. What is a *compensated* op amp? (Is the 741 compensated?)

5. Could we use the DC Sweep method to determine voltage gain if the amplifier were not a DC amplifier?

6. Does the DC low-frequency voltage gain of Figure 20.3 approximately match that calculated by step 6?

Yes   No

21

The Non-Inverting Configuration

VCVS and VCIS

Objective

- *To determine the characteristics of the VCVS and VCIS op amp configurations*

Discussion

A common form of negative feedback is the VCVS (voltage-controlled voltage source) mode of Figure 21.1.

> A voltage-controlled voltage source (VCVS) is a device in which the output *voltage* is directly proportional to the input *voltage*, regardless of the load.

It is useful to think of the VCVS configuration as a *voltage-to-voltage transducer*; that is, there is a linear one-to-one correspondence between the input voltage and the output voltage regardless of load value. In other words, for every input voltage, there is a corresponding output voltage—the load (or output current) does not matter.

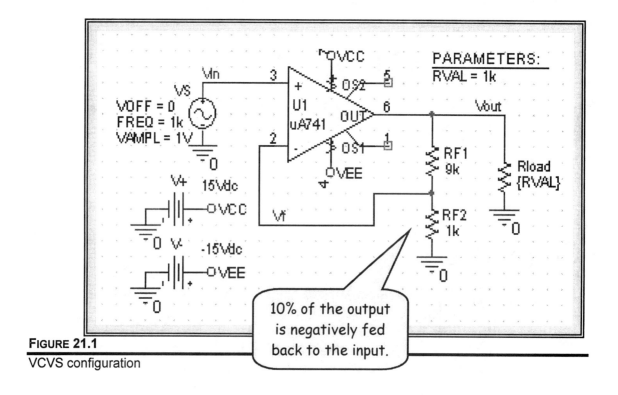

FIGURE 21.1

VCVS configuration

Transfer Function

The *transfer function* is the ratio of output over input. For the VCVS configuration of Figure 21.1, both the input and output are voltage and, therefore, the transfer function is its *voltage gain (A)*. Of special importance to the designer, the VCVS transfer function depends primarily on the feedback resistors (*RF1* and *RF2*).

To calculate the closed-loop transfer function (A_{CL}) for the VCVS amplifier, we refer to the Thevenized version of Figure 21.2. We begin by writing the following equations (where ol = *open loop* and cl = *closed loop*):

$$Vout = Aol \times (Vin - Vf)$$

$$Vf = Vout \times \left(\frac{RF2}{RF1 + RF2} \right)$$

Solving these equations simultaneously yields:

$$Acl = \frac{Vout}{Vin} = \frac{1}{\dfrac{1}{Aol} + \beta} = \frac{Aol}{1 + Aol\beta}$$

β = feedback ratio
= RF2/(RF1 + RF2)

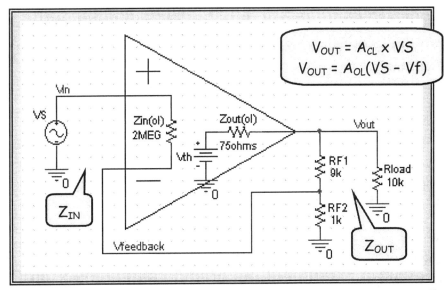

$V_{OUT} = A_{CL} \times VS$
$V_{OUT} = A_{OL}(VS - Vf)$

FIGURE 21.2

Thevenized VCVS circuit

Ideal Case

When performing theoretical calculations in the closed-loop (CL) case, a great simplification results if we assume the *ideal* case for the open-loop (OL) op amp specifications. *We will find that the errors which result from basing our calculations on the ideal case are tiny.*

The table below compares the 741 op amp published specifications for the "big three" parameters with their ideal counterparts.

Open-Loop Parameters

Item	Specification Sheet	Ideal
A_{OL}	200,000	∞
Zin_{OL}	2MEG	∞
$Zout_{OL}$	75ohms	0

Ideal Transfer Function

As an example of the use of an ideal parameter, we substitute the ideal value for open-loop voltage gain (∞) into the equation developed earlier and generate the *ideal* closed-loop voltage gain (transfer function):

$$A_{CL} = \frac{1}{\dfrac{1}{A_{OL}} + \beta} = \frac{1}{\dfrac{1}{\infty} + \beta} = \frac{1}{\beta} = 1 + \frac{RF1}{RF2}$$

Perfect Amplifier

When all the characteristics of the VCVS configuration are put together, we have a nearly perfect voltage amplifier. It gives us ease of design, DC or AC operation, high Z_{IN}, low Z_{OUT}, high bandwidth, and low distortion. It is called *non-inverting* because V_{IN} and V_{OUT} are in phase.

Shadow Rule

During normal operation, negative feedback assures us that the inverting input will track (follow) the non-inverting input. This very useful analysis and troubleshooting aid will be known as the "shadow rule" (named after the famous vaudeville act "Me and My Shadow").

The shadow rule is a direct consequence of negative feedback. *In the ideal case*, the op amp will always generate an output voltage that will drive the input differential (error) voltage to zero. *Therefore, the voltage at the inverting input must track (shadow) the voltage at the non-inverting input.*

VCIS Configuration

The other major form of the non-inverting configuration is the VCIS (*voltage-controlled current source*) of Figure 21.3.

> A voltage-controlled current source (VCIS) is a device in which the output *current* is directly proportional to the input *voltage*, regardless of the load.

As before, think of the VCIS configuration as a *voltage-to-current transducer*. There is a linear one-to-one correspondence between the input voltage and the output current, regardless of load value. For an input voltage, there is a corresponding output current —the load (or output voltage) does not matter.

For this VCIS configuration, the transfer function is I_{OUT}/V_{IN}, which gives units of *transconductance*. Making use of the shadow rule and assuming the ideal case, the closed-loop transfer function is calculated as follows:

$$V_{IN} = V_F \qquad I_{OUT} = \frac{VF}{RF}$$

$$\text{transconductance} = \frac{I_{OUT}}{V_{IN}} = \frac{1}{RF}$$

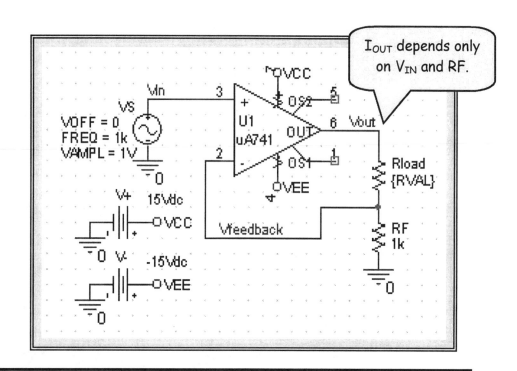

FIGURE 21.3

The VCIS
non-inverting
configuration

Because the load resistor (*Rload*) is not referenced to ground, the VCIS configuration is seldom used. (It will appear in this chapter only as an advanced activity.)

Simulation Practice

Activity *VCVS*

Activity *VCVS* uses the circuit of Figure 21.1 to reveal the characteristics of the VCVS configuration.

1. Create project *noninverting* with schematic *VCVS*.

2. Draw the VCVS circuit of Figure 21.1. (To draw the power supply connections, see *Simulation Note 21.1*.)

Simulation Note 21.1

How do I simplify power connections?

Place Power

To simplify power connections (as shown in Figure 21.1), first move the DC power sources (VDC) to any convenient location on the schematic. Click the *Place Power* toolbar button, place power part *VCC/CAPSYM* at the desired locations, and change the attributes (VCC) as necessary to match the supplies.

3. Set the simulation profile for *Transient* from 0 to 2ms, with a step ceiling of 2μs. Set a *parametric* analysis for global variable *Rload* (parameter name *RVAL*) from 1kΩ to 5kΩ in increments of 1kΩ.

4. To prove that the circuit of Figure 21.1 is a *voltage-controlled voltage source* (VCVS), consider the following:

> For a sine wave input voltage, a circuit is a VCVS if the output voltage is a perfect sine wave whose amplitude is independent of the load (*Rload*).

5. Use parametric analysis to generate the graph of Figure 21.4, which shows V_{IN} and V_{OUT} for five linear values of *Rload* from 1kΩ to 5kΩ.

6. Does the graph of Figure 21.4 seem to prove that the circuit is indeed a near-perfect VCVS? (Does it appear that V_{OUT} is a perfect sine wave whose amplitude is independent of *Rload*?)

        Yes   No

7. Zoom in on any portion of the V_{OUT} curve and note that there are actually five closely spaced curves. Is it now fair to say that the VCVS circuit is a *very good* (but not *perfect*) VCVS? (*Zoom Fit* to return to the original curves.)

Zoom Fit

        Yes   No

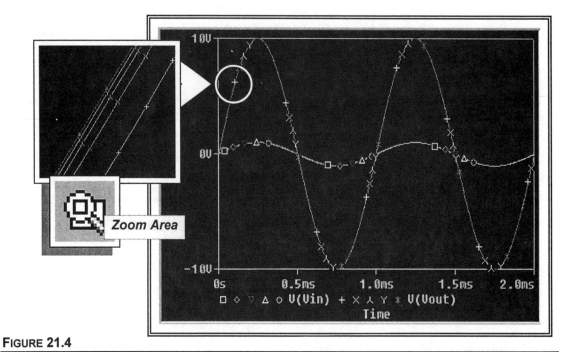

FIGURE 21.4
VCVS test
results

Voltage Gain (Transfer Function)

From this point on, disable the parametric analysis. (*Rload* will automatically revert to 1kΩ.)

8. Determine the voltage gain (*A[CL]*) for each case below.

a. <u>The calculated case</u> (assume *A[OL]* = 200,000).

$$Acl = \frac{1}{\dfrac{1}{Aol} + \beta} = \underline{\hspace{2cm}}$$

b. <u>Using PSpice</u>.

$$Acl = \frac{Vout}{Vin} = \underline{\hspace{2cm}}$$

Input Impedance

9. Determine the input impedance $[Z_{IN}(CL)]$ for each case below. (The equation shown below for $Z_{IN}(CL)$ came from an analysis of Figure 21.2.)

    a. The calculated case [assume $Z_{IN}(OL) = 2MEG$].

    $$ZIN(CL) = (1 + \beta AOL) \times ZIN(OL) = \underline{\hspace{4cm}}$$

    b. Using PSpice. [Add $I(VS)$ to Figure 21.4, and be sure to use peak-to-peak values in your calculations.]

    $$ZIN(CL) = VIN / IIN = \underline{\hspace{4cm}}$$

Output Impedance

10. Determine the closed-loop output impedance $[Z_{OUT}(CL)]$ for each case below.

    a. The calculated case. [Assume $Z_{OUT}(OL) = 75\Omega$.]

    $$ZOUT(CL) = \frac{ZOUT(OL)}{1 + AOL\beta} = \underline{\hspace{4cm}}$$

    b. Using PSpice. (Apply a transient V_{IN} of 1V, measure V_{OUT} with high and low loads—such as 300Ω and 300MEG—and use algebra to determine Z_{OUT}.)

    $$ZOUT(CL) = \underline{\hspace{4cm}}$$

11. To contrast the closed-loop big three characteristics, use previous results to complete the following table:

	A	**Z$_{IN}$**	**Z$_{OUT}$**
Calculated			
PSpice			
Ideal	10	∞	0

12. Based on the results of step 11:

    a. Approximately what degree of error (percentage) do we make in each characteristic by using the ideal case?

    b. Are we justified in using the ideal case for the big three closed-loop characteristics?

         Yes   No

Bandwidth

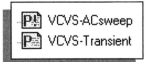

13. Add simulation profile *AC Sweep* from 1Hz to 10MEGHz, 100 points/decade, and add attribute *AC = 1V* to voltage source *VS*. (Be sure to make *ACsweep* the active profile.)

14. Determine the closed-loop bandwidth (BW_{CL}) for each case below.

    a. <u>The calculated case</u>. The bandwidth equals the frequency at which A_{CL} breaks (drops 3dB). As determined below, this happens when A_{OL} drops to 27.65dB.

$$When\ ACL = 7.07\ (70.7\%\ of\ 10\ ) = \frac{AOL}{1 + AOL(1/10)}$$

$$AOL = 24.129\ (27.65dB)$$

        From Figure 21.5, obtain the approximate frequency when A_{OL} is 27.65dB and fill in below. (This is the calculated closed-loop bandwidth!)

        BW_{CL} (calculated) = _____

    b. <u>Using PSpice</u>. Sweep from 1Hz to 10MEGHz, plot *DB(V(Vout)/V(Vin))* and generate the closed-loop curve of Figure 21.6.

        BW_{CL} (PSpice) = _____

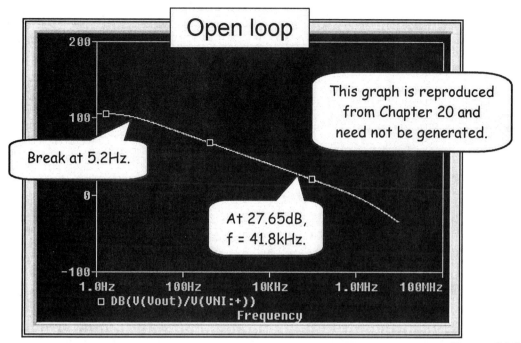

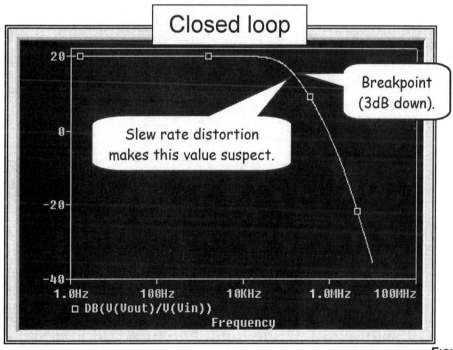

15. Fill in the table below with the calculated and PSpice values of the bandwidth. Do the results agree within 25%? (If not, could slew rate distortion be a major problem?)

Yes   No

	Bandwidth
Calculated	
PSpice	

High-Frequency Effects

So far, all VCVS characteristics have been measured at the low 1kHz frequency (well below the ≈50kHz calculated break frequency). What happens to the beneficial effects of negative feedback as the frequency rises beyond the break frequency?

16. First, fill in the table below for the 1kHz case from step 11.

	A	Z_{IN}
1kHz		
100kHz		

17. Next, increase the frequency to 100kHz (well above the calculated break frequency), and fill in the 100kHz row. (If the output signal clips, lower V_{IN}.)

18. As an additional optional (but very revealing) step, repeat the comparison (1kHz versus 100kHz) for harmonic distortion. (Add a third column to the table and fill it in.)

19. Based on the results of step 16, is it fair to say that the beneficial effects of negative feedback disappear as the frequency increases beyond the breakpoint?

    Yes   No

Shadow Rule

20. Plot graphs of V_{IN} and V_F. Does the inverting input *approximately* track the non-inverting input?   (Zoom in on the overlapping curves.)

    Yes   No

VCVS Summary

21. Based on all the information presented so far, how would you define and describe the VCVS op amp configuration of Figure 21.1?   (Why should its use be restricted to audio frequency applications?)

Advanced Activities

22. Determine how negative feedback affects SNR (signal-to-noise ratio.)

23. For the VCVS configuration, return to parametric analysis and generate the family of Bode plot curves (AC Sweep) of Figure 21.7 by sweeping *RF1* over a range of values (such as 1k, 9k, and 99k).

a. For all curves does the ideal formula for midband gain (1 + RF1/RF2) approximately hold?

Yes   No

b. For all curves is the gain/bandwidth product *approximately* a constant? If so, what is this value?

$$A_{CL} \times BW_{CL} = \underline{\hspace{4cm}}$$

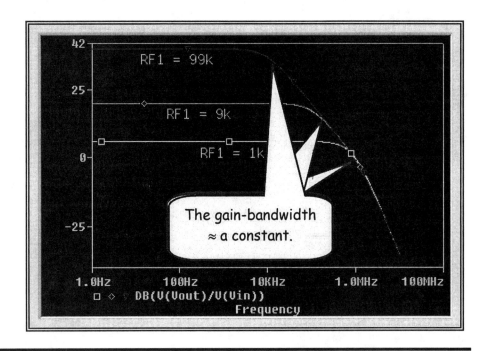

FIGURE 21.7

Family of Bode
plot curves

24. Perform a complete analysis of the VCIS (*voltage-controlled current source*) of Figure 21.3. Determine any or all of the following and compare to the VCVS of Figure 21.1:

a. Graph of $I_{OUT}(CL)$ versus $V_{IN}(CL)$ for various values of RL from 1kΩ to 10kΩ   (Is the circuit a VCIS?)

b. The transfer function (transconductance)

c. $Z_{IN}(CL)$, $Z_{OUT}(CL)$, HD(CL), and BW(CL)

25. Compare the results of step 24 with the ideal characteristics of a VCIS. Would we be justified in using the ideal case for most applications?

Exercises

1. Investigate the output properties of the *buffered* VCVS configuration of Figure 21.8.

2. Investigate the properties of the ideal diode of Figure 21.9.

3. Perform a worst case analysis on the amplifier of Figure 21.1. Which resistor is most critical? (See Chapter 29.)

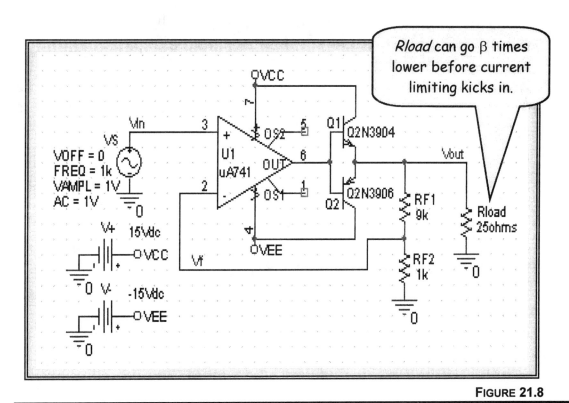

FIGURE 21.8

Buffered VCVS

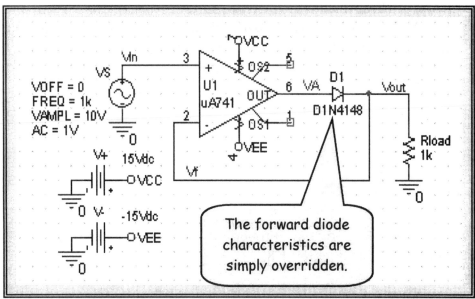

FIGURE 21.9

Ideal diode

Questions and Problems

1. Based on this and the previous two chapters, how would you summarize the effects of negative feedback on VCVS amplifier characteristics?   (How does frequency affect these characteristics?)

2. What is the difference between *open loop* and *closed loop*?

3. For the VCVS circuit, why is there no low-frequency breakpoint?

4. Referring to the circuit of Figure 21.1, approximately what percent error in voltage gain results from assuming the ideal case?

5. Why is the VCVS configuration called *non-inverting*?

6. Based on the results of this chapter:

    a. What is the ideal Z_{IN} for a voltage-controlled (VC) input?

    b. What is the ideal Z_{OUT} for a voltage source (*VS*)?

22

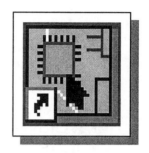

The
Inverting
Configuration

ICVS and ICIS

Objective

- *To determine the characteristics of the ICVS and ICIS op amp configurations*

Discussion

The four basic closed-loop configurations are listed below. The first two were the topic of the previous chapter; the last two are the subject of this chapter.

- VCVS   (voltage-controlled voltage source)
- VCIS   (voltage-controlled current source)
- ICVS   (current-controlled voltage source)
- ICIS   (current-controlled current source)

Because of the lessons we learned with the first two configurations, we can more quickly cover the remaining two. For example, we already know the following:

- The ideal case can be used for most applications with very little error.
- Ideal transfer functions are easy to develop when using the shadow rule.
- A voltage-controlled (VC) input has an ideal Z_{IN} of infinity, and a current-controlled (IC) input has an ideal Z_{IN} of zero.
- A voltage source (VS) has an ideal Z_{OUT} of 0, and a current source (IS) has an ideal Z_{OUT} of infinity.

ICVS Configuration

The most versatile negative feedback configuration for an op amp is the ICVS (current-controlled voltage source) of Figure 22.1.

> A current-controlled voltage source is a device in which the output *voltage* is directly proportional to the input *current*, regardless of the load.

Think of the ICVS configuration as a *current-to-voltage transducer*, with a linear one-to-one correspondence between the input current and the output voltage, regardless of load value. For an input current, there is a corresponding output voltage—the load does not matter.

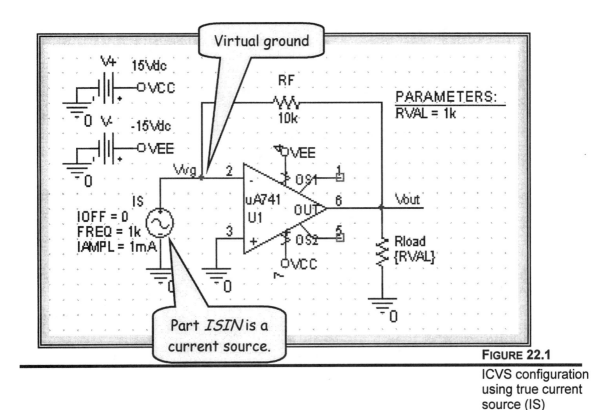

FIGURE 22.1

ICVS configuration
using true current
source (IS)

Virtual Ground

The great versatility of the ICVS configuration is closely related to
the creation of the *virtual ground* at the inverting input. (A *virtual
ground* is at zero volts, but it is not directly connected to ground.)

To see why a virtual ground is created, remember the *shadow
rule*: The inverting input will equal (shadow) the non-inverting input.
Since the non-inverting input is at zero volts (grounded), the
inverting input must also be driven to zero volts (ideally).

ICVS Transfer Function

For the ICVS configuration of Figure 22.1, the input is current and
the output is voltage and, therefore, the transfer function is its
transresistance. In the ideal case, the transfer function depends only
on the single feedback resistor (*RF*). (This time, it will be the
student's responsibility to determine the transresistance equation.)

Handy Current Source

Because the inverting input is at zero volts (a *virtual ground*), we can substitute a voltage source/resistor combination for a true current source (as shown by Figure 22.2). This configuration is called *inverting* because *Vout* is 180° out of phase with V_{IN}.

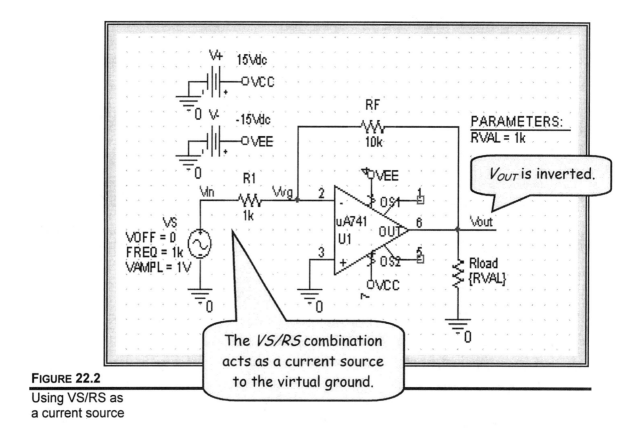

FIGURE 22.2

Using VS/RS as
a current source

If the input is taken as *VS*, the configuration of Figure 22.2 is often called a voltage amplifier (although it does not have the desirable high Z_{IN} characteristics of the VCVS configuration of the last chapter).

ICIS Configuration

The other major form of the inverting configuration is the ICIS (*current-controlled current source*) of Figure 22.3.

A current-controlled current source is a device in which the output *current* is directly proportional to the input *current*, regardless of the load.

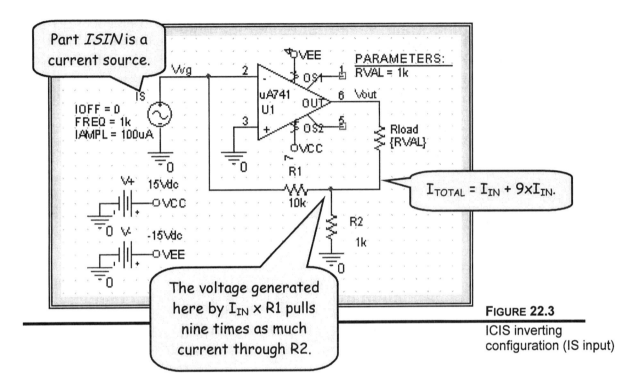

Part *ISIN* is a current source.

IS
IOFF = 0
FREQ = 1k
IAMPL = 100uA

PARAMETERS:
RVAL = 1k

$I_{TOTAL} = I_{IN} + 9 \times I_{IN}$.

The voltage generated here by $I_{IN} \times R1$ pulls nine times as much current through R2.

FIGURE 22.3

ICIS inverting configuration (IS input)

As before, think of the ICIS configuration as a *current-to-current transducer*, with a linear one-to-one correspondence between the input current and the output current, *regardless of load value.* For an input current, there is a corresponding output current—the load does not matter.

ICIS Transfer Function

For the ICIS configuration of Figure 22.3, both the input and output are current and. therefore, the transfer function is *current gain* (β). In the ideal case, the transfer function depends only on feedback resistors (*RF1* and *RF2*). As with the VCIS configuration of the last chapter, the ICIS configuration is seldom used because voltage (not current) is the preferred output variable.

Simulation Practice

Schematic *ICVS*

Schematic ICVS uses the circuit of Figure 22.1 to investigate the characteristics of the *current-controlled voltage source* (ICVS) configuration.

1. Using the ideal case on Figure 22.1, derive an ICVS *transresistance* equation for V_{OUT} versus I_{IN} in terms of *RF*. (*Hint*: Use the shadow rule.)

$$V_{OUT} = (\qquad\qquad\qquad) \times I_{IN}$$

2. Using the equation you developed in step 1, calculate the transresistance gain for the ICVS circuit of Figure 22.1.

$$\text{Transresistance} = V_{OUT}/I_{IN} = \underline{\qquad\qquad\qquad}$$

3. Create project *inverting* with schematic *ICVS*.

4. Draw the ICVS circuit shown in Figure 22.1.

5. Set the simulation profile to *Transient* from 0 to 2ms, with a step ceiling of 2µs. Set the *Parametric* mode to *Global*, variable *RVAL* from 1kΩ to 5kΩ with increments of 1kΩ.

6. To prove that the circuit of Figure 22.1 is an ICVS, consider the following:

> For a sine wave input current, a circuit is an ICVS if the output voltage is a perfect sine wave whose amplitude is independent of the load (*Rload*).

7. Generate the transient graphs of Figure 22.4, which show I_{IN} and V_{OUT} for various values of *Rload* from 1kΩ to 5kΩ.

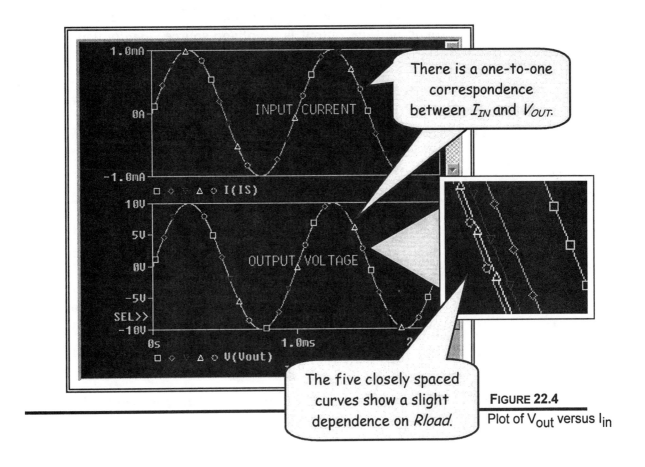

There is a one-to-one correspondence between I_{IN} and V_{OUT}.

INPUT CURRENT

OUTPUT VOLTAGE

The five closely spaced curves show a slight dependence on *Rload*.

FIGURE 22.4

Plot of V_{out} versus I_{in}

8. Based on the waveforms of Figure 22.4, determine the transfer function value (transresistance) as accurately as you can.

    Transresistance = _____

    Is this experimental value (PSpice) close to the calculated value of step 2?

    Yes   No

9. Zoom in on any portion of V_{OUT} and note the five closely spaced curves. Is it fair to say that the circuit of Figure 22.1 is a very good (but not perfect) ICVS with a *transresistance* of 10kΩ?

    Yes   No

10. If the input of the ICVS circuit of Figure 22.1 is a virtual ground and the output is a voltage source, *ideally* what would you expect the circuit's Z_{IN} and Z_{OUT} to be?

Z_{IN} (ideal) = _____ Z_{OUT} (ideal) = _____

11. Measure Z_{IN} and Z_{OUT} using PSpice. (From this step forward, disable the parametric analysis, making *Rload* 10k.)

> *Reminder:* $Z_{IN} = V_{VG}/I(IS)$. To measure Z_{OUT}, determine V_{OUT} with and without a load. (Watch for signal clipping.)

Z_{IN}(PSpice) = _____ Z_{OUT}(PSpice) = _____

12. Comparing steps 10 and 11, are we justified in assuming ideal characteristics for the op amp?

     Yes   No

13. Is the virtual ground at *approximately* zero volts (less than 20mV)?

     Yes   No

Inverting Voltage Amplifier

14. Modify your circuit to generate the amplifier configuration of Figure 22.2, which uses a voltage source/resistor combination for a current source. (Disable the parametric analysis if you wish.)

15. Determine the circuit's voltage gain (V_{OUT}/V_{IN}) using:

a. The calculated ideal case

    A(Ideal) = RF/RS = _____

b. Using PSpice (transient mode)

    A(PSpice) = _____

c. Are the ideal and experimental (PSpice) values approximately the same?

   Yes   No

d. Are the input and output voltages (V_{IN} and V_{OUT}) 180° out of phase?

   Yes   No

16. Using AC analysis determine the bandwidth of the ICVS configuration of either Figure 22.1 or Figure 22.2. (Be sure to set the AC attribute for *VSIN*.)   (Is the amplifier a DC amplifier?)

   Bandwidth = _____

Advanced Activities

17. Determine one or more characteristics of the ICVS configuration (Figure 22.1 or Figure 22.2) before and after the break frequency.   (Does the magic of negative feedback disappear after the break frequency?)

18. Perform a complete analysis of the ICIS (*current-controlled current source*) of Figure 22.3. Determine any or all of the following and compare them to the ICVS of Figure 22.1.

   ▪ I_{OUT} versus I_{IN} for various RL ▪ The transfer function (β)

   ▪ Z_{IN} and Z_{OUT} ▪ Harmonic distortion (HD)

   ▪ Bandwidth (BW)

19. Compare the results of step 18 with the ideal characteristics of an ICIS. Would we be justified in using the ideal case for most applications?

20. Show how to use the ICVS configuration to convert the input current from the collector of a transistor (a current source) to output voltage.

21. As we learned in Chapter 20 (step 21), slew rate distortion involving the 741 op amp sets in at about 8kHz for a voltage swing of 10V.

    For the popular ICVS version of Figure 22.2, view the output waveform at 10V and determine the HD (harmonic distortion) at 5kHz and 10kHz. Repeat for an output voltage amplitude of 5V. Summarize your conclusions. (Be sure to properly adjust the *center frequency* in the Transient dialog box.)

Exercises

1. Draw the *summing* amplifier of Figure 22.5, and show that the output voltage waveform is the algebraic sum of the two input voltage waveforms.

   The 3k resistor (R8) is used to counter the effects of the *input bias current*. The problem arises when this current passes through unequal resistance in the inverting and non-inverting input circuits—resulting in an unwanted differential voltage input. By making the resistances approximately equal (3k $\approx$ 10k∥10k∥10k), this differential voltage is near zero.

2. Determine the transfer function of the *difference amplifier* of Figure 22.6, and test your equation with PSpice.

3. For the *logarithmic amplifier* of Figure 22.7, what is the relationship between V_{OUT} and V_{IN}? [*Hint*: Is the plot of V_{OUT} versus $LOG(V_{IN})$ a straight line?]

4. Predict the output of the ideal clamper of Figure 22.8. Is it a positive or negative clamper? (Change V_{REF} to +5V and predict V_{OUT}.)

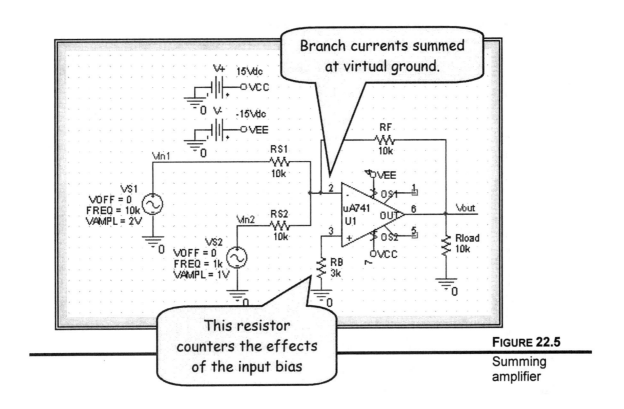

FIGURE 22.5

Summing amplifier

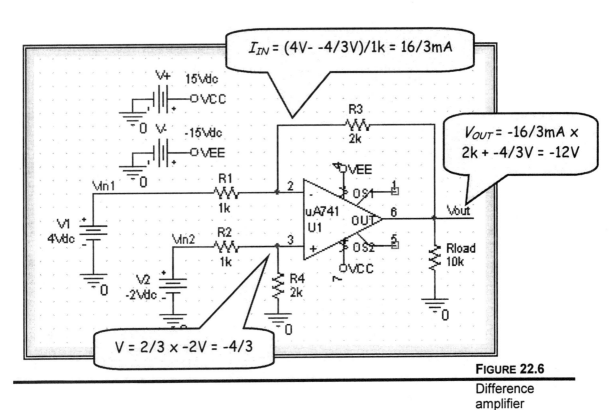

FIGURE 22.6

Difference amplifier

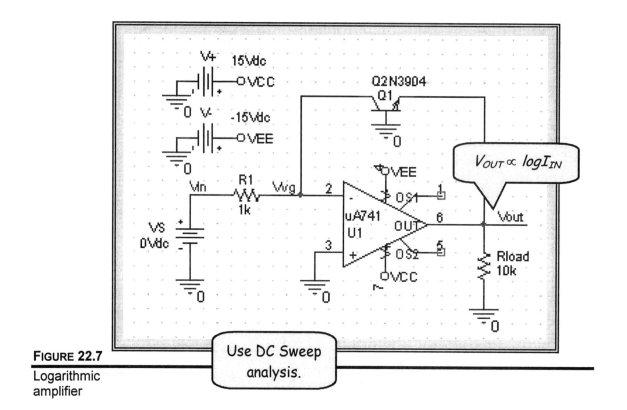

FIGURE 22.7

Logarithmic
amplifier

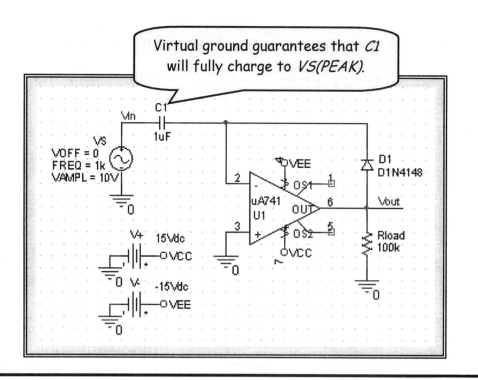

FIGURE 22.8

Ideal clamper

Questions and Problems

1. What is the difference between *ground* and *virtual ground*?

2. When used as a *voltage amplifier*, why is the VCVS configuration of Chapter 21 considered superior to the ICVS configuration (Figure 22.2) of this chapter?

3. A certain ICVS amplifier has a gain/bandwidth product of 10MEG.   If the gain is 80, what is the bandwidth?

4. Regarding the summing amplifier of Figure 22.5:

    a. Could the circuit be used by an analog computer to perform the *add* process?

    b. How does the summing amplifier eliminate *crosstalk* (the circuit of V_{S1} affecting the circuit of V_{S2} and vice versa)?

5.   Place type VCVS, VCIS, ICVS, or ICIS after each of the following:

Z_{IN}	Z_{OUT}	Type
0	0	
0	infinity	
infinity	0	
infinity	infinity	

23

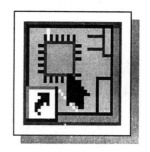

Op Amp Integrator/ Differentiator

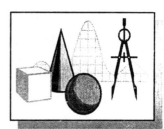

Calculus

Objectives

- *To review the fundamentals of calculus*
- *To perform integration and differentiation using both RC and op amp circuits*

Discussion

We are well aware of the relationships between distance (D), velocity (V), and time (t):

$$D = V \times t \qquad\qquad V = D/t$$

However, the well known *multiply* and *divide* operators ($\times$ and /) only work if velocity is a *constant*.   If velocity is a *variable* (a changing function of time), we must turn to the techniques of *calculus*, developed in the sixteenth century by Isaac Newton. We must *integrate* and *differentiate*.

$$D(t) = \int_0 v(t)dt \qquad V(t) = \frac{dD(t)}{dt}$$

Integration   **Differentiation**

As shown in Figure 23.1, integration is performed graphically by taking the accumulated area under the curve, and differentiation is performed graphically by taking the slope of the curve.

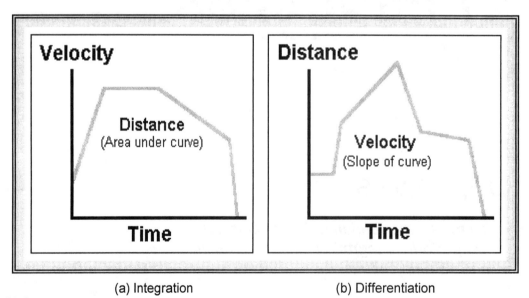

(a) Integration   (b) Differentiation

FIGURE 23.1

Graphical methods of integration and differentiation

Electronic Integration and Differentiation

To see how integration and differentiation can be performed electronically, we note the relationship between current and charge:

$$Q(t) = \int_0^t I(t)dt \qquad\qquad I(t) = \frac{dQ(t)}{dt}$$

However, it is more convenient if both input and output can be measured in voltage. Therefore, we use a resistor as a current-to-voltage transducer ($V = IR$), and we take advantage of the relationship between capacitor voltage and charge ($Q = CV$) to generate:

$$Vout\,(t) = \frac{1}{RC} \int_0^t Vin(t)dt \qquad\qquad Vout\,(t) = RC\,\frac{dVin\,(t)}{dt}$$

Based on these equations, we construct the simple RC-based integrator and differentiator circuits of Figure 23.2.

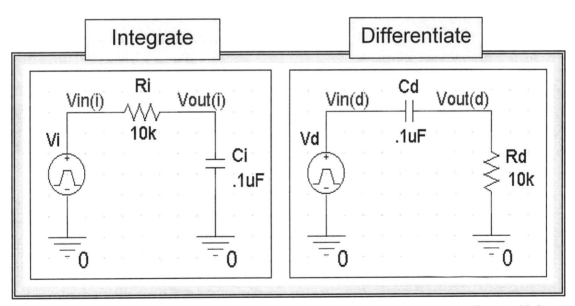

FIGURE 23.2

RC integrator and differentiator

Unfortunately the simple RC-based circuits of Figure 23.2 do not perform perfect mathematical integration and differentiation because the voltage in the capacitor feeds back and affects the current in the resistor. Therefore, to carry out mathematically correct integration and differentiation, we must uncouple the resistor from the capacitor by maintaining the junction at zero volts (a *virtual ground*).

Adding op amps to create the virtual grounds, we arrive at the circuits of Figures 23.3 and 23.4.

Simulation Practice

Activity *INTEGRATOR*

Activity *INTEGRATOR* uses the circuit of Figure 23.3 to perform mathematical integration.

1. Create project *calculus* with schematic *INTEGRATOR*.

2. Draw the op amp integrator shown in Figure 23.3.

3. Set the simulation profile to *Transient*, from 0 to 8ms, with a ceiling of 8μs. Program *VPULSE* (VS) to generate the input waveform of Figure 23.5. The rise and fall times (*td* and *tr*) can be any small value, such as 1ns. (We initialize the capacitor to zero so the output waveforms will be predictable.)

4. Predict the output waveform and sketch your answer on the graph of Figure 23.5. (*Hint*: The output amplitude profile starts at zero and is the inverted accumulated area under the curve of the input waveform.)

5. Generate the output waveform using PSpice, and add your curve to Figure 23.5. Were your predictions correct?

     Yes   No

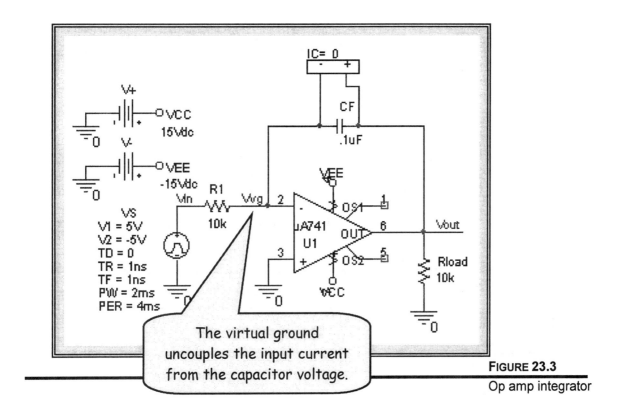

The virtual ground uncouples the input current from the capacitor voltage.

FIGURE 23.3

Op amp integrator

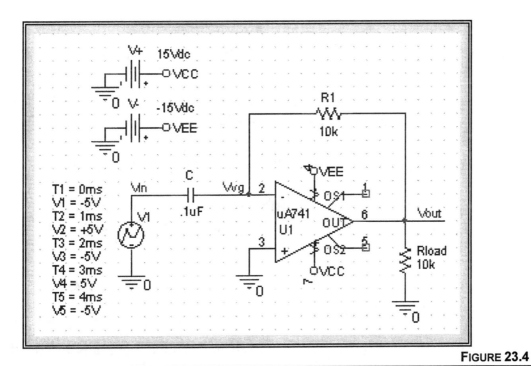

FIGURE 23.4

Op amp differentiator

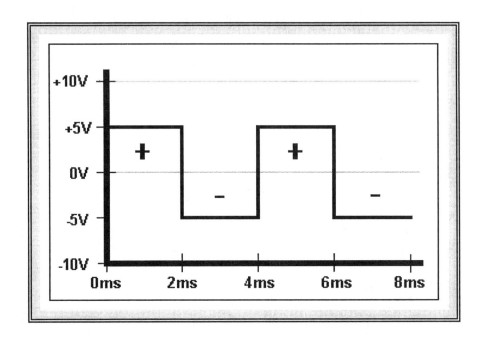

FIGURE 23.5

Integrator input
waveforms

Activity *DIFFERENTIATOR*

Activity *DIFFERENTIATOR* uses the circuit of Figure 23.4 to perform mathematical differentiation.

6. Add schematic *DIFFERENTIATOR* to project *calculus*.

7. Draw the op amp differentiator shown in Figure 23.4, and program *VPWL* (VS) to generate the input waveform of Figure 23.6.

8. Set the simulation profile to *Transient* from 0 to 4ms, with a ceiling of 4µs.

9. Predict the output waveform and sketch your answer on the differentiator graph of Figure 23.6. (*Hint*: The output waveform equals the inverted slope of the input.)

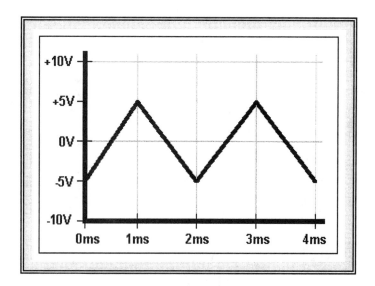

FIGURE 23.6
Differentiator
input waveforms

10. Generate the output waveform using PSpice. (Be prepared for an unexpected result.)

11. It is clear that the system breaks into oscillation due to positive feedback. (At approximately 10kHz, the capacitor gives 180° and the op amp another 180°—and the square wave input does contain this frequency component).

    To solve the problem, add a small damping resistor to the input circuit (Figure 23.7).

12. Again generate the output waveform using PSpice and add your curve to Figure 23.6. Has the oscillation been damped out, and is the new curve similar to your predictions?

    Yes   No

Advanced Activities

13. Using various waveforms, compare the output of the simple RC integrator and differentiator of Figure 23.2 with the op amp counterparts of Figures 23.3 and 23.4. How do they differ?

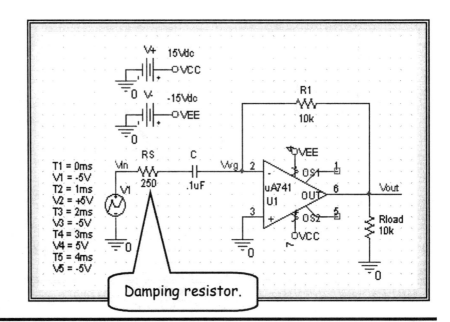

Damping resistor.

FIGURE 23.7

Differentiator with
damping

14. For the input waveforms shown in Figure 23.8, predict the integrated and differentiated output waveforms and draw them on the graph. Using PSpice, generate both the integrator and differentiator waveforms. Did your predictions match the actual case?

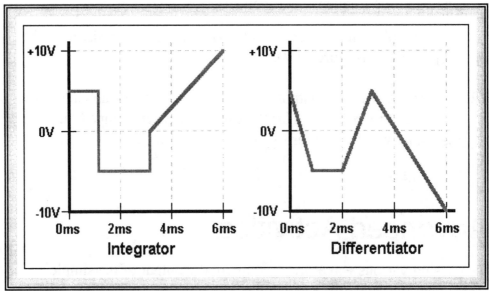

FIGURE 23.8

Input waveforms

15. The compensated integrator of Figure 23.9 uses resistor RF to prevent DC buildup (perhaps due to an offset voltage). Test the circuit over an extended range of pulse repetition rates. Compare your results to the uncompensated integrator of Figure 23.3.

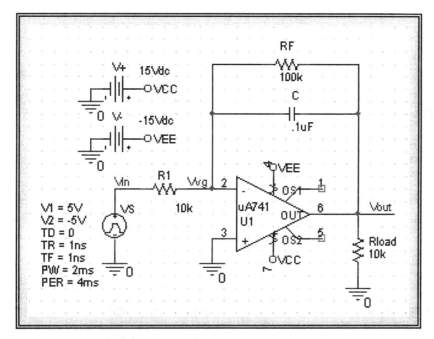

FIGURE 23.9

Compensated integrator

Exercises

1. By cascading two differentiators, design a *double differentiator*. If the input is distance, what is the output? Also, test your design on various input waveforms.

2. Making use of the circuit of Figure 23.10, test a moon-landing simulator which solves the equation that follows for an object of unity mass under control of thrust (T) and gravity (G).

   Using the transient mode, plot various trajectories. (Change the gravity value, and experiment with various VSIN, VPULSE, and VPWL for Vthrust.)

$$D(t) = \iint (T(t) - G(t)) dt^2$$

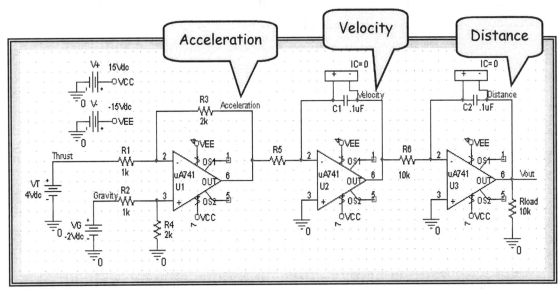

FIGURE 23.10

Moon-landing
simulator

Questions and Problems

1. What is the difference between *multiplication* and *integration*?   Under what
    conditions does integration reduce to multiplication?

2. What is the difference between *division* and *differentiation*?   Under what
    conditions does differentiation reduce to division?

3. When integration is depicted graphically, the output is the area under the curve of the input. Using this fact as a guide, how could you perform integration using an accurate scale and a pair of scissors?

4. Referring to the simple RC integrator of Figure 23.2, why is V_{OUT} equal to the integral of V_{IN} only when the capacitor is nearly empty of charge?

5. Mathematically speaking, under what input conditions would the output of an integrated waveform go up when the input goes down?

6. In real life the differentiator circuit of Figure 23.4 amplifies noise. (The PSpice voltage source is noiseless.)   How does resistor RS of Figure 23.7 reduce noise output (as well as damping out oscillations)?

7. Explain how to determine distance from velocity by using the mechanical integrator of Figure 23.11.

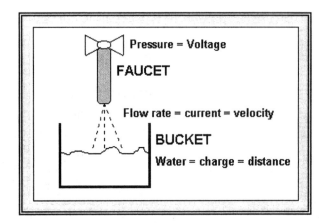

FIGURE 23.11

Mechanical
integration

8. Mathematically speaking, under what input conditions would the output of a differentiated waveform be negative when the input was positive?

24

Oscillators

Positive Feedback

Objectives

- *To determine the conditions necessary for oscillation*
- *To design and test sine and square wave oscillators*

Discussion

Switching from negative to positive feedback requires only a tiny modification in circuit configuration—but a modification that yields an enormous change in circuit operation.

For example, if our body temperature regulation mechanism changed from the stabilizing effects of negative feedback to the runaway effects of positive feedback, we would quickly succumb to a rapidly decreasing or increasing body temperature. When positive feedback is combined with negative feedback, the result is often oscillation.

Specifically, three conditions are necessary for oscillation:

1. Closed-loop gain ≥ 1.

2. Positive feedback (loop phase shift = 360°).

3. Initial spark (noise).

The two major classifications of oscillators are *sine* and *square*. Within each category are numerous varieties and types. This chapter will highlight several popular oscillator configurations.

Simulation Practice

Activity *PHASESHIFT*

Activity *PHASESHIFT* uses the RC network circuit of Figure 24.1 to generate a sine wave.

1. Create project *oscillator* with schematic *PHASESHIFT*.

2. Draw the *phase-shift* oscillator of Figure 24.1, a good choice for low-frequency applications.

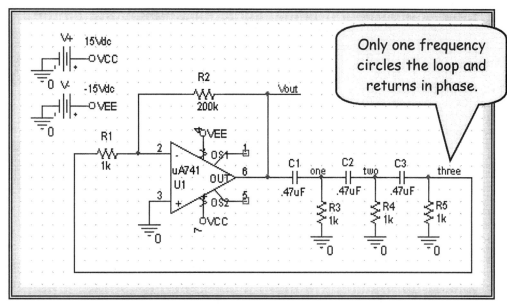

FIGURE 24.1

Phase-shift
oscillator

3. Set the simulation mode to *Transient* from 0s to 300ms, with a ceiling of 100μs.

4. To provide positive feedback, we must have 360° of phase shift. The op amp supplies 180°, so each RC network must supply an average of 60°. To determine the frequency that results in a 60° phase shift, we note the following:

$$\tan 60° = 1.73 = X_C/R \qquad \text{where } X_C = 1/(2\pi fC)$$

Using this equation, determine the predicted frequency.

f = _____

5. Run PSpice and generate the graph of Figure 24.2.

a. Measure the oscillator frequency (1/period) and record the result below:

Frequency of oscillation = _____

PSpice for Windows

b. Is the measured frequency of oscillation *approximately* equal (within 50%) to the calculated value of step 2?

     Yes   No

c. Why does the amplitude increase during the early cycles?

d. What limits the steady-state amplitude?

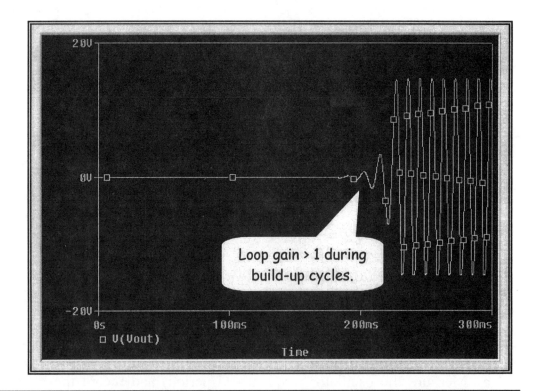

FIGURE 24.2

Phase shift
oscillator output

6. Referring to Figure 24.1, display the waveforms at points *one*, *two*, and *three*.   (Be sure to label the wire segments *one*, *two*, and *three* as shown.)

    a. Do the waveforms show the 60° phase shift?

        Yes   No

    b. Does the amplitude drop as we move from point *one* to point *three*?

        Yes   No

7. Use Fourier analysis to generate the frequency spectrum of Figure 24.3.

    a. Is there a dominant fundamental frequency?

        Yes   No

    b. Does the fundamental frequency approximately equal the time-domain frequency measured in step 5?

        Yes   No

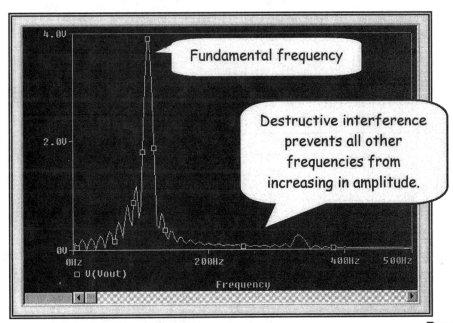

FIGURE 24.3

Phase-shift oscillator frequency spectrum

Activity *COLPITTS*

Activity *COLPITTS* uses the resonant feedback circuit of Figure 24.4 to generate high-frequency sine waves.

8.   Add schematic *COLPITTS* to project *oscillator*.

9.   Draw the *Colpitts oscillator* of Figure 24.4, a good choice for higher frequencies.

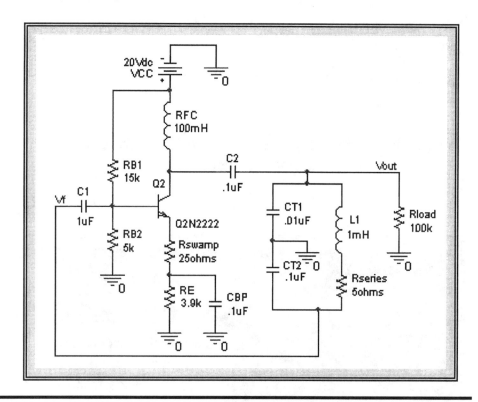

FIGURE 24.4

Colpitts
oscillator

10.   Using the equation below, determine the resonant frequency of the tank circuit, which sets the oscillation frequency.

$$fr = \frac{1}{2\pi\sqrt{LC_T}} = \underline{\hspace{2cm}} \qquad \text{where } C_T = \frac{CT1 \times CT2}{CT1 + CT2}$$

11. Perform a transient analysis to 500µs, and display the output waveform of Figure 24.5.

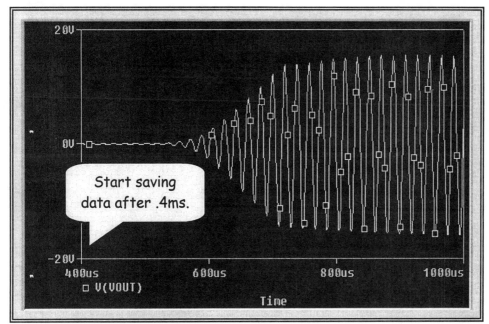

FIGURE 24.5

Colpitts oscillator output waveforms

12. Measure the oscillator frequency (1/period) and compare it to the calculated value of step 10. Are they similar?

    $fr(PSpice) = $ _____

13. Add a plot of Vf (the feedback signal shown in Figure 24.4). Is Vf 180° out of phase with Vout?

    Yes   No

14. Generate a Fourier spectrum for $Vout$ of the Colpitts oscillator.

    a. Is there slightly less distortion (a sharper fundamental frequency with a narrower bandwidth) than with the phase-shift oscillator?

        Yes   No

PSpice for Windows

b. Does the fundamental frequency approximately equal the calculated and measured values of steps 10 and 12?

Yes   No

Activity *SQUARE*

Activity *SQUARE* uses the feedback circuit of Figure 24.6 to generate low-frequency square waves.

---

15. To predict the oscillation frequency of Figure 24.6, we can reason as follows: The signal at the inverting input oscillates between −5V and +5V. Analysis of an RC-pulsed waveform tells us that this requires approximately .7RC time constants. There are two RC swings per cycle. Therefore:

$$f = 1/(2 \times .7 \times RC) = 1/(2 \times .7 \times 10K \times .1\mu F) = \underline{\ 714Hz\ }$$

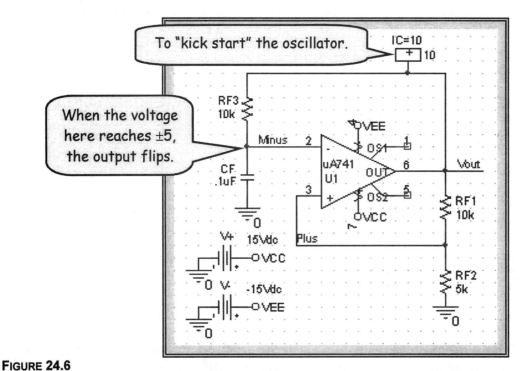

FIGURE 24.6

Square-wave
oscillator

16. Add schematic SQUARE to project oscillator, and draw the square wave oscillator of Figure 24.6, which is good for audio frequencies.

17. Set the simulation profile to *Transient* from 0 to 4ms, and generate the transient output waveforms of Figure 24.7.

    a. Measure the oscillation frequency. How does it compare to the predicted value of step 15?

    $$f \text{ (measured)} = \underline{\hspace{3cm}}$$

    b. Does the circuit "flip" when V(Minus) equals ±5V?

          Yes   No

    c. During switching does the output rise and fall at the slew rate?

          Yes   No

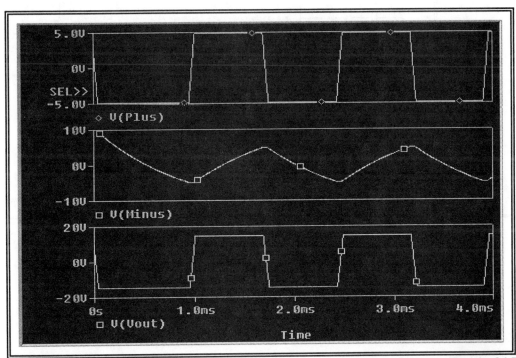

FIGURE 24.7

Square wave waveforms

Advanced Activities

18. The circuit of Figure 24.8 is a Hartley oscillator. It is similar to a Colpitts oscillator but uses tapped inductors instead of tapped capacitors. Using the resonant frequency equation below, design the circuit for an oscillation frequency of 1MEGHz.

$$fr = \frac{1}{2\pi\sqrt{LC}}$$

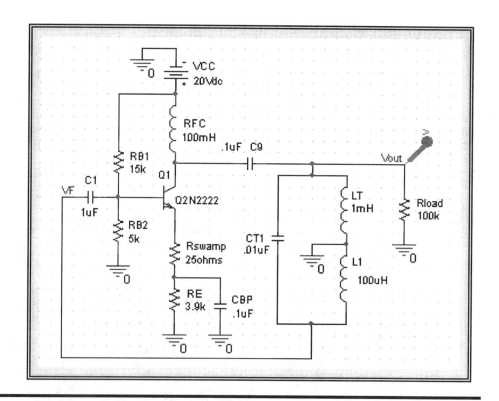

FIGURE 24.8

Hartley
oscillator

19. Repeat step 18 for a Clapp or Armstrong oscillator. (Research will be required.)

20. For any of the circuits of this chapter, experimentally determine the highest oscillation frequency possible.

Exercises

1. By cascading a square wave oscillator with an op amp integrator, design and test a triangle wave generator.

2. Design a Colpitts oscillator using an FET instead of a bipolar transistor.

Questions and Problems

1. What are the three conditions necessary for oscillation?

2. During what period of oscillator operation is the closed-loop gain greater than 1? (*Hint*: How does the signal initially build up?)

3. For a real-world oscillator, what serves as the spark?

4. For the Colpitts oscillator of Figure 24.4, explain how the loop phase shift is 360º.

25

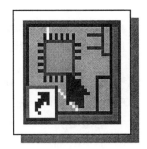

Filters

Passive and Active

Objectives

- *To contrast passive and active filters*
- *To compare low-pass, high-pass, and bandpass filters*

Discussion

A *filter* is a device in which the gain is designed to be a special function of the frequency. The major classifications of filters are:

- Low-pass — passes low frequencies.
- High-pass — passes high frequencies.
- Bandpass — passes frequencies within a band.
- Band-stop (notch) — blocks frequencies within a band.

Any of these filters can be either passive or active. A passive filter uses only resistors, capacitors, and inductors. An active filter usually is implemented with an op amp that employs both positive and negative feedback. A *first-order* filter contains a single RC or RL network and rolls off at 20dB/dec; a *second-order* filter contains two RC or RL networks and rolls off at 40dB/dec.

Passive Filter

A passive first-order, low-pass filter is shown in Figure 25.1. To solve such a filter circuit, let's use complex numbers to determine its break frequency and rolloff.

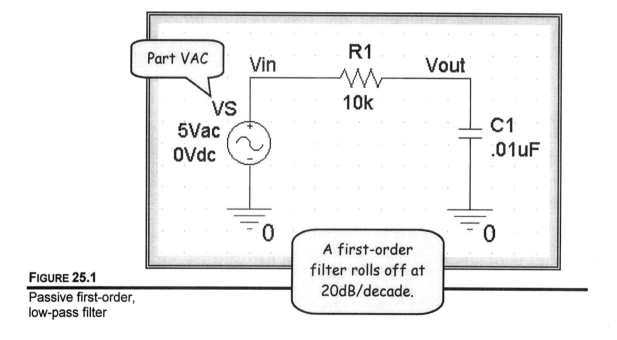

FIGURE 25.1

Passive first-order, low-pass filter

We start with the filter's transfer function in rectangular form:

$$H(j\omega) = \frac{V_{OUT}}{V_{IN}}(j\omega) = \frac{-jX_C}{R - jX_C} = \frac{1}{1 + j\omega RC} \quad \text{where } \omega = 2\pi f$$

Switching to polar form yields the magnitude and angle:

$$H = \frac{1}{\sqrt{1 + (\omega RC)^2}} \angle - \tan^{-1}(\omega RC)$$

By definition, the break (critical) frequency occurs when the angle equals 45° (or the magnitude drops by 3dB).

$$\tan(45°) = \omega RC = 1 \quad \omega_C = \frac{1}{RC} \text{ or } f_C = \frac{1}{2\pi RC}$$

Therefore, $f_B = \dfrac{1}{2\pi RC} = \dfrac{1}{2\pi \times 10k \times .01\mu F} = 1{,}591.5 Hz$

To determine rolloff, note that at high frequencies (well beyond the breakpoint):

$$\omega RC \gg 1 \quad \text{So, the magnitude} \cong \frac{1}{\omega RC} = \frac{1}{2\pi f RC} \propto f^{-1}$$

Therefore, if f increases by a factor of 10 (a decade), then:

$$Rolloff = 20\log_{10} 10^{-1} = -20 dB / decade$$

Active Filter

Many variations of active filters exist. One of the most popular is the VCVS (*voltage-controlled voltage source*) *Sailen-Key network* configuration of Figure 25.2—popular because R and C network components are of equal value. Complex mathematical analysis yields a break frequency the same as the passive case (1/2πRC = 1,591.5Hz), but a rolloff that doubles to 40dB/decade.

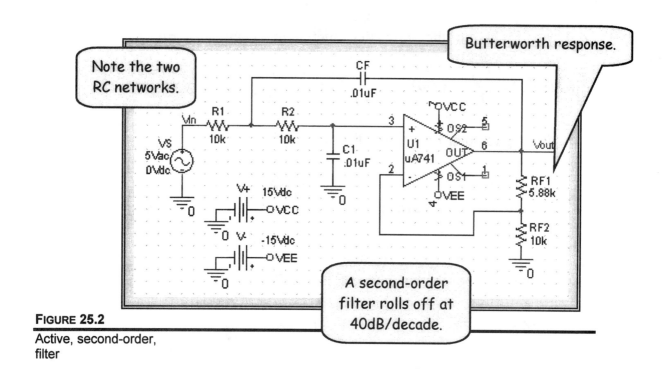

FIGURE 25.2

Active, second-order, filter

One advantage of using an active filter is that it offers gain. In fact, its characteristics depend on gain. For example, changing the gain of the active filter circuit of Figure 25.2 changes its response, as shown in Table 25.1. The Butterworth response is especially attractive because it gives the flattest response.

Response	RF1/RF2	Breakpoint	Gain
Bessel	0.288	1.27fc	1.27 (2.1dB)
Butterworth	0.588	1.00fc	1.588 (4.0dB)
1dB Chebyshev	0.955	0.863fc	1.955 (5.8dB)
2dB Chebyshev	1.105	0.852fc	2.105 (6.5dB)
3dB Chebyshev	1.233	0.841fc	2.233 (7.0dB)

TABLE 25.1

Major filter types

Simulation Practice

Activity *LOWPASSIVE*

Activity *LOWPASSIVE* uses the circuit of Figure 25.1 to test the characteristics of a passive, low-pass RC filter.

1. Create project *filters* with schematic *LOWPASSIVE*.

2. Draw the first-order, passive, low-pass filter of Figure 25.1.

3. Set the simulation profile for AC Sweep from 1Hz to 100MEGHz, with 100 points/decade.

4. Run PSpice and generate the Bode plot of Figure 25.3.

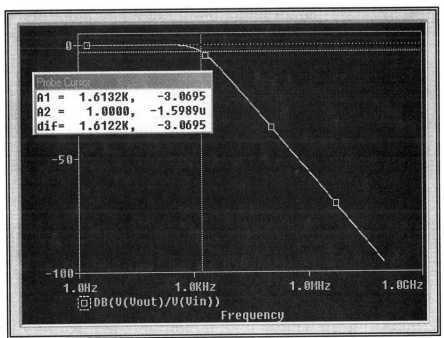

FIGURE 25.3

Passive filter
Bode plot

5. From Figure 25.3, complete the following table with the PSpice values.   Are they approximately the same?

	Calculations	PSpice
F_{BREAK}	1,592Hz	
Rolloff	20dB/decade	

Activity LOWACTIVE

Activity *LOWACTIVE* uses the circuit of Figure 25.2 to investigate the properties of an active low-pass filter.

6. Add schematic *LOWACTIVE* to project *filters*.

7. Draw the active low-pass filter of Figure 25.2.

8. Set the simulation profile to AC Sweep from 1Hz to 100MEGHz, with 100 points/decade.

9. Run PSpice and append the results to the passive plot. (**File, Append Waveform, CLICKL** on desired file, **OPEN, Do not skip sections**).   The result is shown in Figure 25.4.

10. Looking at the curves:

   a. Do both curves break at 1,591Hz?

   Yes   No

   b. Does the active filter rolloff at twice the rate of the passive filter (20dB/decade versus 40dB/decade)?

   Yes   No

   c. Does the active filter show a small ($\approx$4dB) gain?

   Yes   No

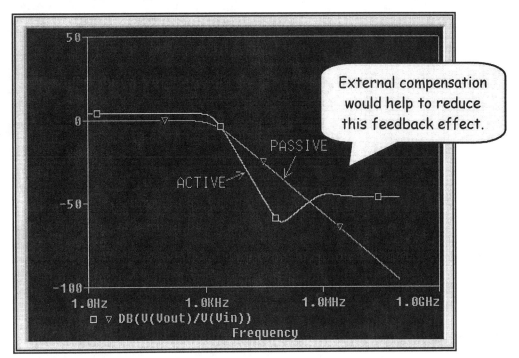

FIGURE 25.4

Active/passive
comparison

11. Return to the active filter of Figure 25.2 and sweep *RF1/RF2* over the values shown in Table 25.1. Display curves for all five types of filters, as shown by Figure 25.5. (*Suggestion*: Keep *RF2* constant and sweep RF1 over the values 2.88k, 5.88k, 9.55k, 11.05k, and 12.33k.)

12. Based on Figure 25.5:

   a. Is the Butterworth filter the smoothest (maximally flat)?

      Yes   No

   b. Does the order of the Chebyshev filters (1dB, 2dB, and 3dB) equal its increase in gain about the breakpoint [that is, does $A_{DB}(peak) - A_{DB}(low\text{-}freq) = 1dB, 2dB, 3dB$]?

      Yes   No

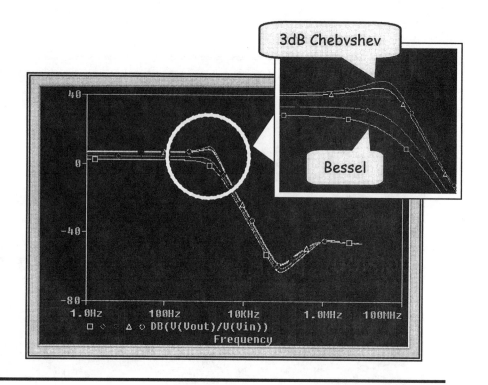

FIGURE 25.5

Comparison of
filter types

Advanced activities

13. Draw the *multiple-feedback bandpass* filter of Figure 25.6.

14. When C1 = CF, the formulas listed below determine the values of the resistors. Solve the equations for a center frequency (f_0) of 1kHz, a Q of 5, and a gain (A_0) of 1. (Assume C1 = CF = .01μF.)

   R1 = $Q/(2\pi f_0 C A_0)$ = _____

   R2 = $Q/(2\pi f_0 C(2Q^2 - A_0))$ = _____

   RF = 2R1

15. Enter the calculated resistor values into the circuit of Figure 25.6, and display the Bode plot of Figure 25.7.

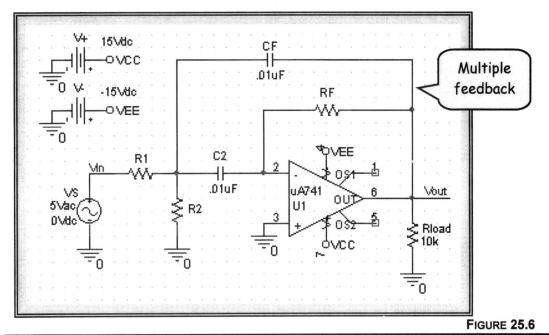

FIGURE 25.6

Bandpass filter

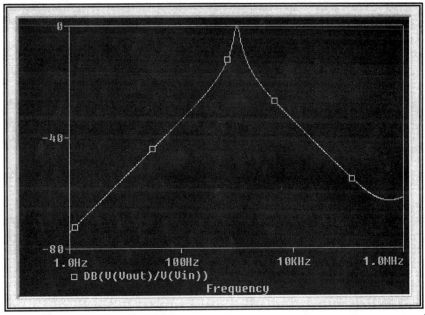

FIGURE 25.7

Bode plot

16. Using the cursors, measure the values for A_0, f_0, and Q (f_0/BW) and fill in Table 25.2. Are the measured and calculated values approximately the same?

Yes   No

	Design goal	Actual
A0		
f0		
Q (f0/BW)		

TABLE 25.2

Bandpass comparison

17. By reversing the Rs and Cs within the RC networks of Figures 25.1, 25.2, and 25.6, design and test passive and active high-pass and band-stop filters of your choice.

18. Substitute an inductor for the resistor in the passive low-pass filter of Figure 25.1. Set L to the value that gives the same breakpoint (when $X_L = R$). Is the filter still a low-pass filter, and is the rolloff still 20dB/decade?

Exercises

1. Using the bandpass filter of Figure 25.6, design a circuit to pass the note of A (880Hz) but reject by at least 10dB the next note below (G#, at 830.61) and the next note above (A#, at 932.33).

2. Cascade two active second-order, low-pass filters (Figure 25.2) and determine their combined characteristics. (How does the breakpoint and rolloff compare to the single-stage configuration?)

Questions and Problems

1. What are several advantages of using an active filter over a passive filter?

2. Why don't filters generally make use of inductors?

3. What order is the bandpass filter of Figure 25.6?   (*Hint*: How many RC networks does the filter have?)

4. Show how a combined (cascaded) low-pass and high-pass filter can be used to construct a bandpass filter.

5. Why is a Butterworth filter preferred for many applications?

26

The Instrumentation Amplifier

Precision CMRR

Objectives

- *To analyze the instrumentation amplifier*
- *To compare the CMRR of an instrumentation amplifier with a single op amp amplifier*

Discussion

Suppose we wish to design a biofeedback system. One approach is to attach two electrodes to the arm a slight distance apart. We then send the amplified differential voltage to a meter, and we close the loop by attempting to mentally control the voltage.

However, this approach confronts us with several technical problems. First of all, the biofeedback voltage is a very small *differential* signal, but the entire arm is immersed in a strong sea of electromagnetic interference. Second, the transducers attached to the arm have high output impedances and, therefore, supply only very small currents. For the feedback process to work, we must amplify the differential-mode information signal and reject the *common-mode* interference signal.

In purely technical terms, we require a high input-impedance differential amplifier of exceptionally high CMRR (common-mode rejection ratio). Also, to match the circuit to different individuals, it would be a real plus if we could easily and quickly control the gain. Such a circuit is called an *instrumentation amplifier* and is shown in Figure 26.1.

Careful mathematical analysis shows that the differential mode (DM) voltage gain of the circuit is:

$$A_{DM} = \frac{2R1 + RG}{RG}$$

Simulation Practice

Activity *IA*

Activity *IA* (Instrumentation Amplifier) uses the circuit of Figure 26.1 to show how three VCVS op amps are arranged to provide high CMRR and variable gain.

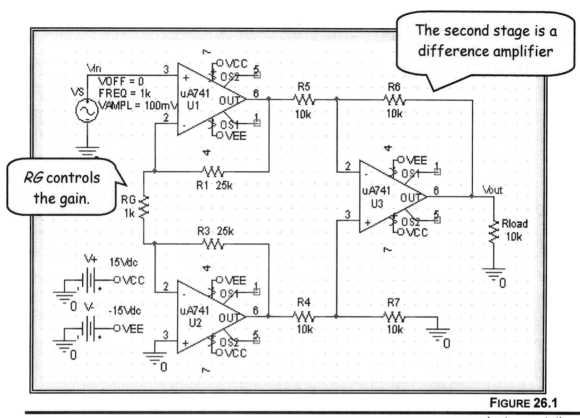

FIGURE 26.1

Instrumentation
amplifier

1.   Create project *operationalamplifier* with schematic *IA*.

2.   Draw the circuit of Figure 26.1, and set the following simulation
     profiles:

     a.   *Transient* from 0 to 2ms, with a ceiling of 2μs.

     b.   *AC Sweep* from 1Hz to 100MEGHz, with 100points/decade.

3.   Using both calculations and PSpice, determine A(DM) when RG
     = 100Ω, 1kΩ, and 10kΩ and place your answers below.

	Calculations	**PSpice**	
A_{DM} at RG = 100Ω	_____	_____	
A_{DM} at RG = 1kΩ	_____	_____	
A_{DM} at RG = 10kΩ	_____	_____	

PSpice for Windows

4. With RG = 1kΩ, use PSpice to determine $Z_{IN}(DM)$.

   $Z_{IN}(DM)$ = _____

5. Tie together the two inputs (pin 3 of the input op amps) and determine the common-mode voltage gain A_{CM}. By combining A_{DM} and A_{CM}, determine the CMRR.

$$CMRR = \frac{ADM}{ACM} = \underline{\hspace{3cm}}$$

6. Based on the results of steps 4 and 5, is the $Z_{IN}(DM)$ and CMRR much higher for an instrumentation amplifier than conventional single, op amp amplifiers?

   Yes   No

7. Generate a Bode plot of the instrumentation amplifier and report the bandwidth:

   Bandwidth = _____

   Is the instrumentation amplifier a DC amplifier?

   Yes   No

Advanced Activities

8. Remove the ground from the right side of R7 (Figure 26.1) and substitute a +5V source. If this node were assigned a pin, what effect would various values of voltage have on Vout?

9. Generate a gain-bandwidth product family of curves for the instrumentation amplifier. Is the gain-bandwidth product a constant?

10. Set resistor tolerances and perform a worst-case analysis. Is an instrumentation amplifier more stable than a conventional op amp amplifier? (See Chapter 29.)

Exercises

1. Assuming a 1μV bio-feedback signal and a .1V common-mode interference signal, test the effectiveness of the instrumentation amplifier as a bio-feedback device. (*Suggestion*: Select different frequencies for the biofeedback and interference signals so they can be distinguished in the output signal.)

2. Substitute a LF411, LM324, or LM111 op amp (from library Eval) and remeasure the CMRRs. How do they compare?

Questions and Problems

1. What is the voltage gain range of the instrumentation amplifier of Figure 26.1?

2. What makes the instrumentation amplifier a DC amplifier?

3. For the amplifier of Figure 26.1, what is the voltage gain of the second (differential) stage (involving resistors R4 through R7 and op amp U3)?

4.   Why do instrumentation amplifiers often interface transducers? (*Hint*: Typically, what is the Z_{OUT} of a transducer?)

5.   On the input of the common mode amplifier of Figure 26.1, the common-mode signal is 10 times as big as the differential signal. What is the ratio on the output?   (*Hint*: Make use of the answer to step 5.)

Part 6

Special Processes

In Part 6 we introduce a number of very special processes available under PSpice. These include *noise analysis*, *Monte Carlo analysis*, *worst-case analysis*, *performance analysis*, the *Histogram*, and *analog behavioral modeling*.

We will find that PSpice "instantly" presents results that would take many hours using hand analysis and calculation on actual circuits.

27

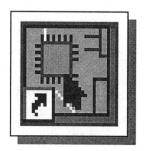

Noise
Analysis

Signal-to-Noise Ratio

Objectives

- *To perform a noise analysis on an amplifier*
- *To determine the signal-to-noise ratio*

Discussion

The noise-generating devices in a circuit are the resistors and the semiconductor devices. A *noise* analysis tells the designer how the noise from all such devices will affect an output signal. Once we know the noise, we can easily generate the *signal-to-noise ratio* at the output node—a term that tells us how important the noise is in relation to the output signal.

Noise analysis can be done only in conjunction with an AC analysis. The noise function generates a noise density spectrum for each device over a range of frequencies and performs an RMS sum at the specified output node. Also reported is the equivalent noise from a specified input source that would cause the same output noise value if injected into a noiseless circuit.

Simulation Practice

Activity *AMPLIFIER*

Activity *AMPLIFIER* will perform a noise analysis on the amplifier of Figure 27.1.

1.   Create project *noise* with schematic *AMPLIFIER*, and draw (or bring back from Chapter 8) the amplifier circuit of Figure 27.1.

2.   Set the simulation profile for *AC Sweep/Noise* and fill in the *Simulation Settings* dialog box as shown in Figure 27.2.

- *Output Voltage* [*V(Vout)*] gives the node at which the AC noise is to be determined.

- *I/V Source* (VS) is the independent voltage or current source at which the equivalent input noise will be calculated.

- *Interval* (100) causes a detailed table to be printed to the output file for every hundredth frequency. (If no value is specified, no tables will be generated.)

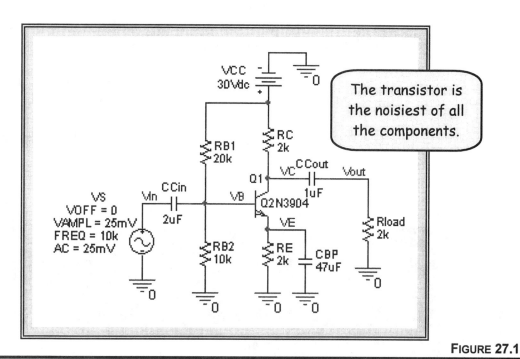

FIGURE 27.1

Amplifier test circuit

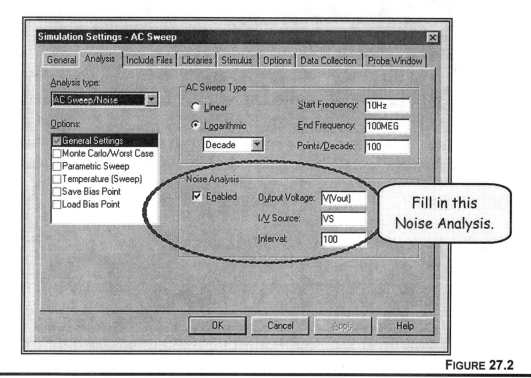

FIGURE 27.2

AC Sweep and Noise Analysis dialog box

3. Run the analysis and generate the *noise density* (volts/root Hz) plots of Figure 27.3. (*Note*: Volts/root Hz = volts/Hz$^{1/2}$.)

    ▪ Y-axis 1 shows *ONOISE* (output noise density)—the RMS summed noise (volts/root Hz) at the output node [*V(Vout)*].

    ▪ Y-axis 2 shows *INOISE* (input noise density)—the equivalent RMS input noise (volts/root Hz) at V_{IN} (V_S).

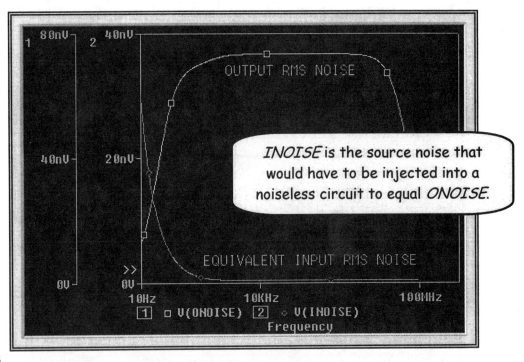

FIGURE 27.3

Noise plots

4. Based on the results (Figure 27.3), what is the output noise density (*ONOISE*) at the midband frequency of 100kHz? At this same frequency, what is the equivalent noise density at the input (*INOISE*)?

    VRMS/Hz(ONOISE) at 100kHz = _____

    VRMS/Hz(INOISE) at 100kHz = _____

5. As shown in Figure 27.4, add a second plot and display the input and output *signal* voltages.

    What is the *signal-to-noise* ratio (SNR) at 100kHz? (Give the ratio of output signal voltage to output noise voltage density in both regular and dB format.)

Signal-to-noise ratio at 100kHz (regular) = _____

Signal-to-noise ratio at 100kHz (dB) = _____

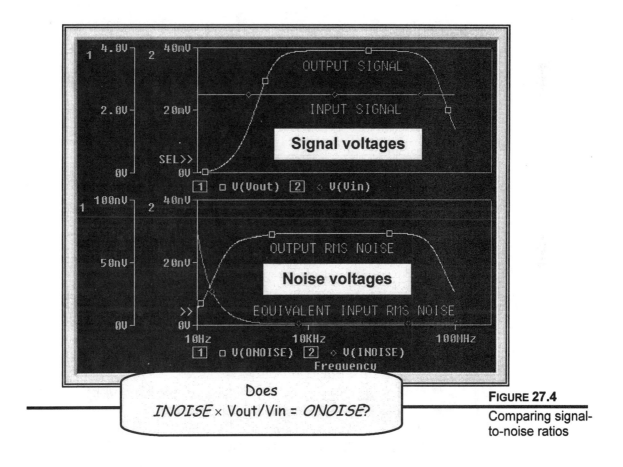

FIGURE 27.4

Comparing signal-to-noise ratios

6. Examine the *noise analysis* section of the output file (**View, Output File**). Per our instructions (Figure 27.2), was a separate noise analysis tabulated in detail for each decade of frequencies from 10Hz to 100MEGHz?

            Yes   No

7. Table 27.1 shows the noise analysis section in the output file for 100kHz.

a. Does most of the noise come from the transistor?

Yes   No

b. Which resistor or resistors contribute the greatest noise?

c. Is the *equivalent input noise at VS* (4.699E–10 V/RT HZ) times the transfer function gain (156.3) equal to the *total output noise voltage* (7.346E–8 V/RT HZ)?

Yes   No

```
****  NOISE ANALYSIS          TEMPERATURE =  27.000 DEG C

              FREQUENCY =  1.000E+05 HZ
****  TRANSISTOR SQUARED NOISE VOLTAGES (SQ V/HZ)
                         Q_Q1
```

RB	4.051E–15
RC	4.259E–23
RE	0.000E+00
IB	2.014E–17
IC	1.310E–15
FN	0.000E+00
TOTAL	5.381E–15

```
****  RESISTOR SQUARED NOISE VOLTAGES (SQ V/HZ)
          R_RC     R_RB1    R_RL     R_RB2     R_RE
TOTAL   7.470E–18 5.130E–22 7.470E–18 1.026E–21 5.134E–23
****  TOTAL OUTPUT NOISE VOLTAGE   = 5.396E–15 SQ V/HZ
                                   = 7.346E–08 V/RT HZ
          TRANSFER FUNCTION VALUE:
          V(Vout)/V_VS          = 1.563E+02
    EQUIVALENT INPUT NOISE AT V_VS = 4.699E–10 V/RT HZ
```

TABLE 27.1

Noise analysis
at 100kHz

Advanced Activities

8. Increase the temperature to 100°C (212°F), and again determine the total noise at 100kHz. (How does it compare to the noise at the default temperature of 27°C?)

   Total V/RT HZ (noise) at 100kHz and 27°C = _7.346E-08_

   Total V/RT HZ (noise) at 100kHz and 100°C = _____

Exercises

1. Perform a noise analysis on the audio amplifier of Figure 15.5. At 100kHz how does the signal-to-noise ratio (dB) compare to that of the single-stage amplifier of this chapter (Figure 27.1)?

2. Perform a separate noise analysis on the bipolar amplifier of Figure 8.1 and the JFET amplifier of Figure 15.1. Based on your results, is a bipolar circuit noisier than a similar JFET amplifier?

Questions and Problems

1. Perform a root-mean-square (RMS) of the numbers below:

   RMS of 7, 2, –4, 8, 12 = _____

2. What causes noise in a resistor?

3. Based on Figure 27.3, what is the noise bandwidth of the amplifier circuit of Figure 27.1?

4. Referring to Figure 27.3, why does the equivalent input RMS noise rise significantly at lower frequencies? (*Hint*: What role do the capacitors play?)

5. Determine the voltage gain of the amplifier of Figure 27.1 using each of the following methods and compare to the *TRANSFER FUNCTION VALUE* (156.3) of Table 27.1.

    a. The ratio of resistors method [$(rc||rl)/re'$, where $re' = 25\text{mV}/I_{EQ}$].

    b. Directly from Figure 27.4 (midband).

    c. Using PSpice in the transient mode.

6. If it turns out that the signal-to-noise ratio of the amplifier of Figure 27.1 is too low (too much noise), what would be the most logical step to increase the ratio? (*Hint*: Referring to Table 27.1, does the transistor give the greatest noise component?)

7.   Referring to Table 27.1, how is 7.346E–08V/RT HZ obtained from 5.396E–15SQ V/HZ?

8.   Based on the results of procedure step 8, why does the total noise voltage go up as the temperature goes down?

9.   Referring to Figure 27.4, does *ONOISE = INOISE × A(Vout/Vin)* at midband (100kHz)?

28

Monte Carlo Analysis

Tolerances

Objectives

- *To assign tolerance values to components*
- *To perform Monte Carlo analysis on a tank circuit*

Discussion

Consider the familiar tank circuit of Figure 28.1. If such a circuit is part of a radio transmitter, then critical characteristics such as impedance and bandwidth must be very carefully controlled in order to avoid trouble with the FCC (Federal Communications Commission).

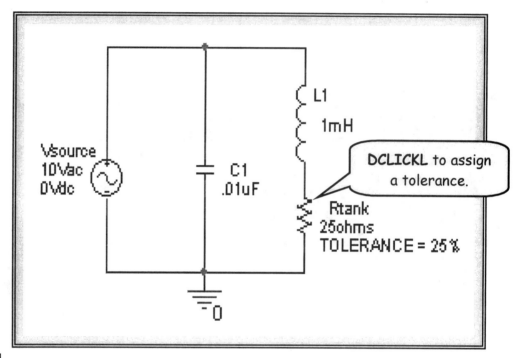

FIGURE 28.1

Tank circuit

Here's where we get into trouble: As the tank circuits come down the assembly line, *Rtank* varies randomly according to its tolerance value.   For example, if its tolerance is 10%, then all we know for sure is that *Rtank's* value lies between $25\pm6.25\Omega$. A good designer must know how these random tolerance variations affect the circuit characteristics.

To provide the answer, PSpice offers *Monte Carlo* analysis.

Monte Carlo

During a Monte Carlo analysis, PSpice performs several runs of a DC, AC, or transient analysis, each time varying component values randomly within the tolerance range. This random nature gives the process its "Monte Carlo" tag. The first run is always the nominal run, using the component's face value, with no tolerance variations.

For simple parts (such as resistors), tolerance values are easily set as attributes; for more complex parts (such as transistors), tolerance values are set within the model definition. Output data is sent to *Probe* for graphical display and to the *output file* for tabular display.

In this chapter, we will first perform a Monte Carlo analysis on the tank circuit of Figure 28.1 with only resistor *Rtank* given a tolerance.

Simulation Practice

Activity *TANK*

This activity will use Monte Carlo analysis to determine how the tolerance of a tank resistor affects the impedance waveform.

1. Create project *montecarlo* with schematic *TANK* and page PAGE1.

2. Draw the circuit of Figure 28.1, and set *Rtank's* tolerance to 25%. (**DCLICKL** on *Rtank's* symbol, set and display property *Tolerance* to 25%.)

3. Bring up the Simulation Settings dialog box and set the simulation profile for a logarithmic AC Sweep from 10k to 100k, at 500 points/decade. Click the *Monte Carlo/Worst Case* option and fill in the analysis tab as shown in Figure 28.2. (Do not click **OK** to close the dialog box as yet.)

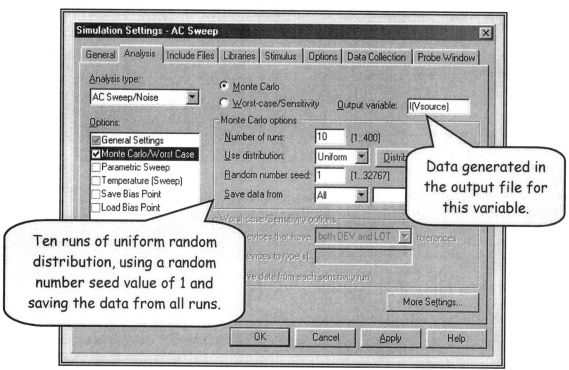

FIGURE 28.2

Monte Carlo
settings

4. Click the *More Settings* button to bring up the *Monte Carlo/Worst-Case Output File Options* of Figure 28.3.

The *output file options* of Figure 28.3 determine the type of data written to the output file. During processing, the system searches each specified waveform [such as *I(Vtank)*] and finds the single point that matches your selection.

- *YMAX* finds the greatest deviation from nominal.

- *MAX* finds the maximum value without regard to nominal.

- *MIN* finds the minimum value without regard to nominal.

- *RISE_EDGE* finds the first positive-going crossing of the zero axis.

- *FALL_EDGE* finds the first negative-going crossing of the zero axis.

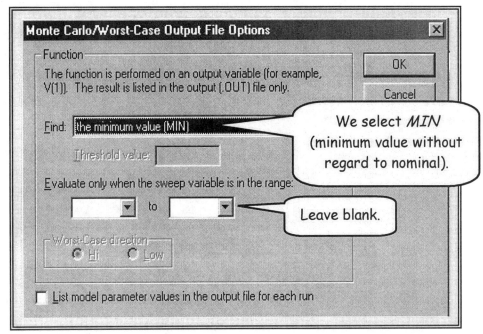

Monte Carlo/Worst-Case Output File Options ⊠

┌─ Function ──────────────────────────────
The function is performed on an output variable (for example,
V(1)). The result is listed in the output (.OUT) file only.

Find: the minimum value (MIN)

Threshold value:

Evaluate only when the sweep variable is in the range:

[▼] to [▼]

┌─ Worst-Case direction ─
  ● Hi   ○ Low

☐ List model parameter values in the output file for each run

OK
Cancel

We select MIN
(minimum value without
regard to nominal).

Leave blank.

FIGURE 28.3

Monte Carlo
options

5. Select the MIN option, **OK**, **OK** when done.

> We select MIN because we are interested in maximum
> impedance, which occurs at minimum current. *Remember*:
> Only currents or voltages can be specified as output variables.
> (*Evaluate only when the sweep variable is in the range* can be
> left blank.)

6. Run PSpice and perform the analysis. When the Available
   Sections dialog box comes up, **OK** to select **ALL** (all runs) by
   default.

7. Display the circuit impedance in dB to generate the random
   family of curves of Figure 28.4. (As shown, be sure to zoom in
   on the peak portion of the curves.)

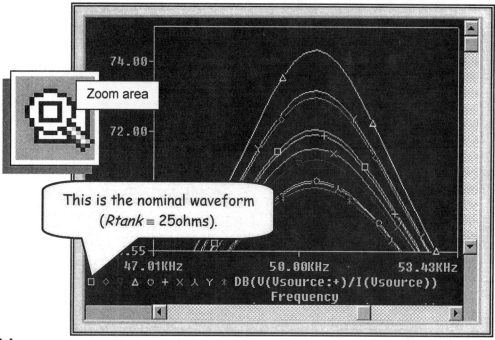

FIGURE 28.4

Monte Carlo
output waveform

8. Looking at the results (Figure 28.4), what is the approximate percent variation between the lowest and highest peak values compared to the average peak value? (How does this compare to the 25% tolerance value?)

$$\% \text{ variation} = \frac{V_{PEAK}(difference)}{V_{PEAK}(average)} \times 100 = \underline{\hspace{2cm}}$$

Average = (high + low)/2.

9. Repeat the whole process using 10% tolerance, and record the variations below:

TOLERANCE	% VARIATION
25%	
10%	

Output File

During a Monte Carlo analysis, various data is written to the output file. Most of the data is related to the output variable specified by Figure 28.2 [*I(Vsource)*] and to the option selected in Figure 28.3 (*MIN*). The most important data is found in the *SORTED DEVIATIONS* section of Table 28.1.

10. Display the output file (**View, Output File**), and scroll to find the *SORTED DEVIATIONS* section (Table 28.1).

    For each run *past the nominal* (2 through 10), the minimum *I(Vsource)* amplitude is found and listed in order of magnitude as an absolute number, a percent of nominal, and a standard deviation (sigma). Also listed is the frequency when the maximum deviation occurred and whether the deviation was lower or higher than nominal.

SORTED DEVIATIONS OF I(V_Vsource) TEMP = 27.00DEG C **MONTE CARLO SUMMARY**		
RUN	**MINIMUM VALUE**	**AT FREQ.**
Pass 10	3.0257E-03 (121.4% of Nominal)	at F = 50.3500E+03
Pass 8	2.9757E-03 (119.39% of Nominal)	at F = 50.3500E+03
Pass 5	2.9551E-03 (118.57% of Nominal)	at F = 50.3500E+03
Pass 3	2.7585E-03 (110.68% of Nominal)	at F = 50.3500E+03
Pass 7	2.6552E-03 (106.53% of Nominal)	at F = 50.3500E+03
Pass 6	2.5347E-03 (101.7% of Nominal)	at F = 50.3500E+03
NOMINAL	2.4924E-03 (100% of Nominal)	at F = 50.3500E+03
Pass 2	2.2423E-03 (89.966% of Nominal)	at F = 50.3500E+03
Pass 9	2.1918E-03 (87.941% of Nominal)	at F = 50.3
Pass 4	1.9129E-06 (76.752% of Nominal)	at F = 50.3500E+03

Minimum value

TABLE 28.1

Minimum value results

PSpice for Windows

11. Based on your *sorted deviations* data, what was the minimum value in *I(Vsource)* and at what frequency did it occur?

    Minimum value = _____ Amps

    Found at = _____ Hz during run _____

Advanced Activities

12. Plot *I(Vsource)* and compare the displayed curves to the *SORTED DEVIATIONS* table of Table 28.1.

13. Rerun the analysis using a *Gaussian* distribution. (Referring to Figure 28.2, select distribution *Gaussian* instead of the default *Uniform*.) Summarize the differences.

Exercises

1. By assigning *Rtank* a 25% tolerance, perform a Monte Carlo analysis on the Class C amplifier of Figure 11.2.

2. By assigning *Cspeedup* a 25% tolerance, perform a Monte Carlo analysis on the bipolar switching circuit of Figure 16.5.

3. Repeat either of the above by switching to a 1% tolerance. Summarize your results.

Questions and Problems

1.   When running a Monte Carlo analysis, exactly what is randomized?

2. Which *output file option* would you use to determine the first positive-going crossing of the zero axis?

3. How would you repeat a Monte Carlo analysis using a different sequence of random parameter values?

4. When doing a Monte Carlo analysis, the first run is always the _____ run.

5. What does "nominal" mean?

6. Using the equation $Z = Q^2 r$, determine the expected change in Z for a 25% change in r. How does your result compare to the Monte Carlo generated values? [*Hint*: Q = X_L(at resonance)/r.]

29

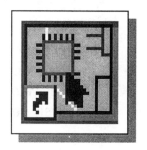

Worst-Case Analysis

AC Sensitivity

Objectives

- *To perform a sensitivity analysis on a tank circuit*
- *To perform a worst-case analysis on a tank circuit*

Discussion

Although both Monte Carlo and worst-case analysis make use of component tolerances and involve a number of runs, they are quite different. Monte Carlo analysis, covered in the last chapter, varies component values in a random manner as the runs are made; worst-case analysis does not use random variations at all.

Instead, worst-case analysis involves a two-step process. First, we perform a *sensitivity* analysis, in which model parameters are varied one at a time for each device, with a DC, AC, or transient analysis run for each variation.

When the sensitivity analysis is done, PSpice uses the corresponding data to perform one final worst case run, with each parameter set up or down by its *full* tolerance in such a way as to produce the greatest output signal or the greatest deviation from nominal (or other result, depending on the function chosen).

Simulation Practice

Schematic *TANK*

We stay with the familiar tank circuit of Figure 29.1 to determine the worst-case tolerance effects.

1. Create project *worstcase*, schematic *TANK*, and page *PAGE1*.

2. Draw the tank circuit of Figure 29.1 (reproduced from Chapter 28), and set the tolerance of all three components to 10%.

3. Bring up the Simulation Settings dialog box:

   • Set the main simulation profile for a logarithmic AC Sweep from 10k to 100k, at 500 points/decade.

- Click and enable the *Monte Carlo/Worst Case* option and fill in this *analysis* tab as shown in Figure 29.2. (Do not click *OK* yet.)

> *Dev* is appropriate for discrete circuits in which devices vary independently. *Lot* is appropriate for integrated circuits, in which all devices vary as a group.

- Click the *More Settings* box and bring up the *Monte Carlo/ Worst Case Output File Options* box of Figure 29.3.

Of the five *Find* options available, we choose *MIN* (the minimum value), which searches for the minimum tank current value [*I(Vtank)*] of each curve. When choosing *MIN*, the *Worst-Case direction* must be Low.

OK, **OK** when done.

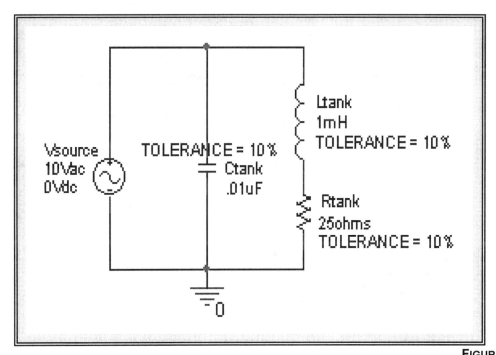

FIGURE 29.1

Tank circuit

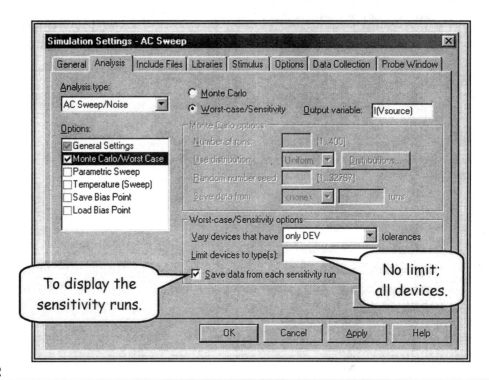

FIGURE 29.2

Worst case
settings

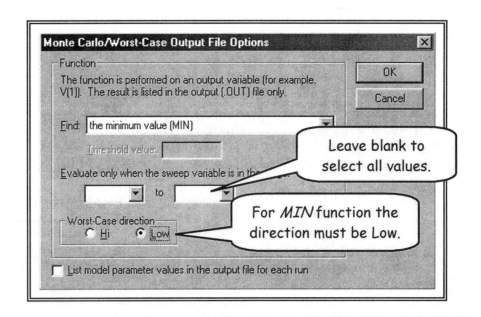

FIGURE 29.3

Output file
options

4. Run PSpice, **OK** (since **All** available sections is selected by default), and access the default output screen.

> *Note*: …MINAL means the nominal waveform, …L DEVICES means the waveform for all devices, and the other options specify waveforms to be generated for individual devices.

5. Add the impedance trace and expand the axes to generate the individual parameter and final (worst case) output waveforms of Figure 29.4.

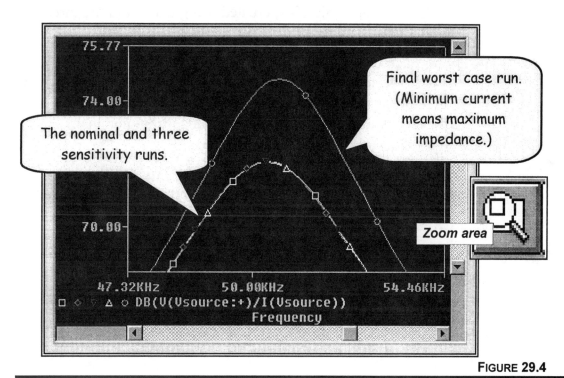

FIGURE 29.4

Worst-case
waveform set

6. Referring to Figure 29.4, note that five waveforms were drawn. In order of generation and listing, they are as follows:

    1 nominal waveform
    3 sensitivity waveforms for each component (R, C, and L)
    1 final worst case waveform

7. If you wish, zoom in on the four closely clustered nominal and sensitivity waveforms and verify that there are actually four *separate* waveforms.

8. Using the cursor, record below the worst-case maximum impedance (minimum current).

$$\text{Maximum } Z = \underline{\hspace{4cm}}$$

Output File

9. Scan through the output file and locate the *SENSITIVITY SUMMARY* of Table 29.1, created using data from the three AC Sweep sensitivity runs.

    Each of the three components is increased one at a time by 1% for each run, and (as directed by the MIN function) the new minimum value is located. The devices are then ranked from the largest to the smallest percent change from nominal.

SORTED DEVIATIONS OF I(V_Vsource) TEMPERATURE = 27.000 DEG C
SENSITIVITY SUMMARY
****

RUN	MINIMUM VALUE
C_Ctank C_Ctank C	2.4954E-03 at F = 50.3500E+03 (1.2124% change per 1% change in Model Parameter)
R_Rtank R_Rtank R	2.4948E-03 at F = 50.3500E+03 (.9938% change per 1% change in Model Parameter)
NOMINAL	2.4924E-03 at F = 50.3500E+03
L_Ltank L_Ltank L	2.4904E-03 at F = 50.3500E+03 (-.7804% change per 1% change in Model Parameter)

TABLE 29.1

Sensitivity summary

> For example, when *Ltank* is increased by 1%, the minimum current decreases by .7804% from nominal (to 2.4904mA).

10. As an example, is it reasonable that when *Rtank* is increased, the minimum current should also increase?

    Yes   No

11. Continue to scan through the output file and locate the *UPDATED MODEL PARAMETERS* data of Table 29.2, which shows how the three components were increased or decreased by their full 10% tolerance during the final worst case run. (1.1 and .9 are scaling factors for the nominal values.)

UPDATED MODEL PARAMETERS   TEMPERATURE = 27.000 DEG
WORST CASE ALL DEVICES

DEVICE	MODEL	PARAMETER	NEW VALUE
C_Ctank	C_Ctank	C	.9 (Decreased)
L_Ltank	L_Ltank	L	1.1 (Increased)
R_Rtank	R_Rtank	R	.9 (Decreased)

.9 = decrease by 10%.
1.1 = increase by 10%.

TABLE 29.2
Worst-case parameter changes

12. Is it reasonable that *Rtank* should be decreased in order to decrease the minimum current?

    Yes   No

13. When all tolerances are set as listed in Table 29.2, the system performs the final worst case run, and the summary data of Table 29.3 are written to the output file.

    a. Is the absolute minimum possible (worst case) current value 1.8371mA and does it occur at 50.35kHz?

       Yes   No

    b. Is the minimum possible current 73.709% of the minimum nominal current?

       Yes   No

c. Does 20log(10V/1.8371mA) equal the maximum impedance value recorded in step 8?

Yes   No

```
SORTED DEVIATIONS OF I(V_Vsource)  TEMPERATURE =   27.000 DEG C
                       WORST CASE SUMMARY
************************************************************************

        RUN                    MINIMUM VALUE

     NOMINAL          2.4924E-03 at F =   50.3500E+03

   ALL DEVICES        1.8371E-03 at F =   50.5820E+03
                        ( 73.709% of Nominal)
```

TABLE 29.3

Worst case
summary

14. For the tank circuit of Figure 29.1 (using 10% tolerance values), the maximum impedance possible is 74.717dB and the minimum possible current is 1.8371mA.

True   False

Advanced Activities

15. Lower the tolerances of all three components to 1% and re-evaluate the system. Summarize the changes.

16. Repeat the worst-case analysis using the *MAX* function (instead of the *MIN* function). Explain the differences. (*Hint*: *MAX* finds the greatest amplitude, without regard to nominal.)

Exercises

1. Perform a *complete* worst-case analysis on the Class C amplifier of Figure 11.2.

2. Perform a *complete* worst-case analysis on the bipolar switching circuit of Figure 16.5.

Questions and Problems

1. Which of the following analyses generate random numbers?

   a. Monte Carlo
   b. Sensitivity
   c. Worst case

2. When performing a worst-case analysis, why must a sensitivity analysis be done first?

3. Based on the sensitivity data of Table 29.1, the tank current is *least* sensitive to which component?

4. Based on Table 29.1, by what percent would *I(Vsource)* change if *Rtank* changed by 20%?

5. To give the minimum (worst case) current, write UP or DOWN before each of the following.

   _____ C _____ L _____ R

6.   What is the lowest (worst case) value of *I(Vsource)* that can ever be expected?

30

Performance Analysis

Goal Functions

Objectives

- *To help design a tank circuit using performance analysis*
- *To introduce the goal function*

Discussion

In Chapter 15 (Volume I) we used parametric analysis on the tank circuit of Figure 30.1 to generate the family of curves of Figure 30.2.

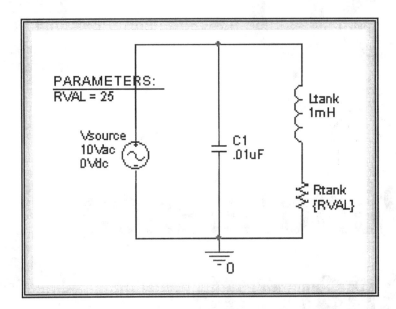

FIGURE 30.1

Initial circuit

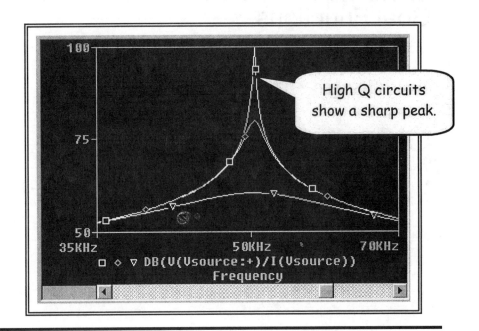

FIGURE 30.2

Family of curves of
various values of Q

Now suppose we are designing the tank circuit for a radio station, and we must know the precise value of *Rtank* that results in a peak Z of exactly 70dB.   To obtain this value from the family of curves of Figure 30.2 is quite difficult.   Chances are that the curve we want lies between several curves, and we are faced with the demanding process of interpolation to approximate the right answer.   There must be a better way.

There is, and it's called *performance analysis*.   With performance analysis we directly generate a graph of maximum Z versus *Rtank*.   We can then pick the correct value right off the curve.   Best of all, it's very easy to implement:

One: Define a variable (such as *RVAL*), and set up a parametric analysis.

Two: Select *Performance Analysis* as an X-axis processing option.

Three: Add a trace expression that includes a *goal function*.

As the system performs each parametric calculation for each value of *Rtank*, it searches for and remembers the coordinates of each peak value of Z.   When done, it then processes the information and generates a graph of peak Z versus *Rtank*.

Simulation Practice

Activity *TANK* (Performance)

Performance analysis will determine the precise value of *Rtank* required to give a maximum impedance of 70dB for the tank circuit of Figure 30.1.

1.   Create project *performance* with schematic *TANK* and page PAGE1.

2.   Draw and save the tank circuit of Figure 30.1.   Be sure to set the *PARAMETERS* component.

3. Set up the simulation profile for a logarithmic *AC Sweep* from 10kHz to 100kHz at 500 points/decade, and a linear parametric sweep of *RVAL* from 10Ω to 100Ω in increments of 1Ω.

4. Run PSpice, accept **All** *Available Settings*, and create the default output screen of Figure 30.3.

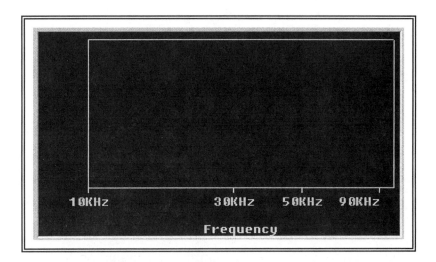

FIGURE 30.3

Initial output plot

5. Click the *Performance Analysis* toolbar button to create the graph of Figure 30.4. Note that the X-axis is changed to the parameter variable (*RVAL*).

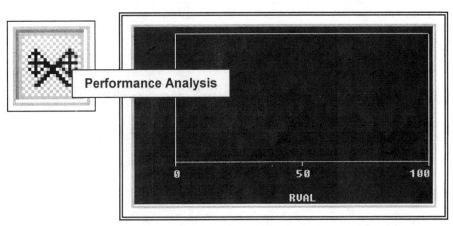

FIGURE 30.4

Setting the X-axis to *Rtank* (RVAL)

6. Click the *Add Traces* toolbar button to bring up the Add Traces dialog box, with goal functions automatically listed. From this list we must choose the desired goal function.

   Since we are seeking the maximum Z value of each curve, we choose (**CLICKL**) goal function *Max(1),* where the "1" indicates the variable to be searched for its maximum value. For us, we enter the circuit impedance, **OK**.

$$Max(dB(V(Vsource:+)/I(Vsource)))$$

7. Looking at the resulting graph (Figure 30.5), we see the desired graph of maximum Z versus *R1*.

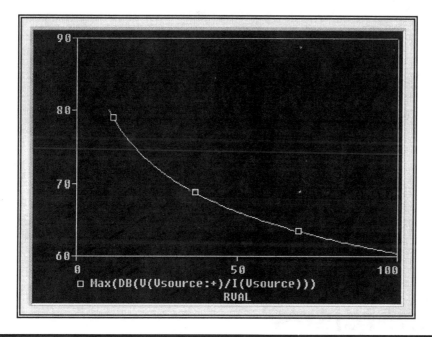

FIGURE 30.5

Final graph of peak Z versus *R1*

8. Using the cursor, pick off the value of RVAL that gives a peak Z of as close as possible to 70dB. (*Note*: For greater accuracy, some interpolation still may be necessary.)

   RVAL (for 70dB) = _____

9. Exit the parametric mode and assign to *Rtank* the value determined from step 8.

10. Run PSpice and determine the maximum value of *Z* using the cursor.

    Max Z = _____

    Is this value very close to 70dB?

    Yes   No

11. If you wish, select any other value of *Rtank* from Figure 30.5 and verify the results. Did the performance analysis graph correctly predict the actual response?

    Yes   No

Advanced Activities

12. Change the increment values of the parametric sweep of *Rtank* from 1Ω to 25Ω. What does the shape of the *Z* versus *Rtank* curve tell us about the trade-off between the number of sampling points and accuracy?

13. Use *the CenterFreq(1,db_level)* goal function to plot a graph of center frequency versus *Rtank*. (Use 3 for db_level.)

Exercises

1. Based on the tank circuit of Figure 30.1, determine and test the value of *Rtank* that gives a bandwidth of 10k. (*Hint*: Use the goal function shown below.)

    Bandwidth(1,db_level)

2. Using both the *Max(1)* and *Min(1)* goal functions, plot a graph of gain versus *Rload* for the bipolar amplifier of Figure 8.1. (*Hint*: *Max(V(Vout))-Min(V(Vout))* will give the peak-to-peak output voltage.)

3. Again, using the tank circuit of Figure 30.1, use performance analysis (plot a graph of frequency versus C) to determine the value of C that gives a center frequency of 50kHz.

Questions and Problems

1. A performance analysis must be accompanied by a parametric analysis.

   True   False

2. Referring to the following goal function, what is the meaning of the "1"?

   Max(1)

3. When conducting a performance analysis, what is the X-axis variable?

4. The return values of a goal function are plotted on the Y-axis.

   True   False

5. How does the user convert the X-axis to the parameter variable?

6. Referring to the graph of Figure 30.5, what is the unit of the Y-axis values?

31

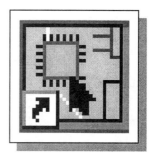

The Histogram

Gaussian Distribution

Objectives

- *To generate a histogram from the results of a Monte Carlo analysis*
- *To contrast linear and Gaussian distributions*

Discussion

Now that we have completed *Monte Carlo* analysis and *performance* analysis, a question naturally arises: Is there any value in combining the two?

The answer is *yes*. As a case in point, consider the familiar tank circuit of Figure 31.1. Monte Carlo analysis gives us a way of varying *Rtank* about its 10% tolerance value. Performance analysis gives us a way of recording the maximum value for each run and displaying the results. When we combine the two we get a *histogram*.

With a histogram the X-axis takes on the goal function values (the maximum impedance) and the Y-axis takes on units of percent. Using histograms we will determine the percentage probability that the tank circuit will lie in various X-axis (maximum impedance) slots due to the random tolerance variations of *Rtank*.

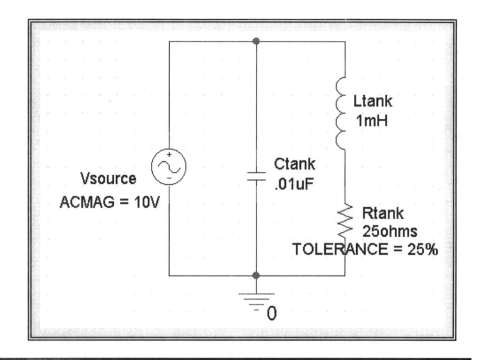

FIGURE 31.1

Tank circuit

Simulation Practice

Schematic *TANK* (Histogram)

The histogram of schematic *TANK* will show how the maximum impedance is distributed as *Rtank* is varied randomly.

1. Create project *histogram* with schematic *TANK* and page PAGE1, and draw the tank circuit of Figure 31.1.

2. Create simulation *AC Sweep*, and set the simulation profile for a logarithmic AC Sweep from 10k to 100k, at 500 points/decade.

    Click the *Monte Carlo/Worst-Case* option and fill in as in Chapter 28, **OK**. [*Output variable* = I(Vtank), *number of runs* = 10, *Use distribution* = uniform, *Random number seed* = 1, *Save data from* = All.]

3. Run PSpice, leave **All** available sections selected by default, **OK**, and generate the initial graph.

4. Switch to *performance analysis* (click the *Performance Analysis* toolbar button) and note the initial histogram chart.

5. Enter the familiar impedance trace expression shown below and generate the impedance histogram of Figure 31.2.

$$Max(DB(V(Vsource:+)/I(Vsource)))$$

6. In your own words, state what is plotted on the X and Y axes.

PSpice for Windows

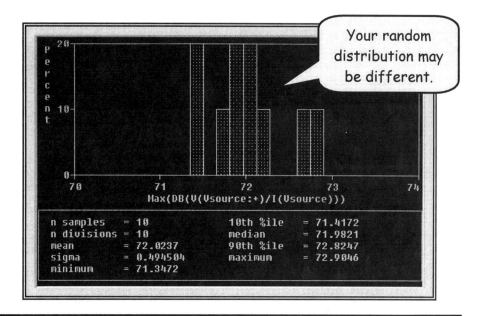

FIGURE 31.2

Impedance
histogram

7. The histogram shows the percentage probability that the circuit impedance will fall into one of 10 impedance slots. By reviewing Figure 31.2, answer the following.

    a. What is the size of each impedance slot?   [*Hint: (Maximum – minimum)/10.*]

        Each slot = _____

    b. What impedance range (or ranges) has the highest probability of occurring?

        Highest probability range = _____

    c. What is the probability that the impedance lies between the 40th and 60th percentile?

        Probability (40th–60th percentile) = _____

    d. Why are the mean (average) and medium (middle) values different?

Advanced Activities

The impedance histogram of Figure 31.2 seems unrealistic. Shouldn't the responses be clustered about the center (have a bell-shaped distribution)?

8. To change the histogram from the default *uniform* case to the more realistic *Gaussian* distribution: Click the *Edit Simulation Settings* toolbar button, **Monte Carlo/Worst-Case** option, under *Use Distribution* change *uniform* to *Gaussian*, **OK**.

9. Run PSpice, **OK** (*All* settings), switch to performance analysis (**Plot**, **Axis Settings**), and enter the impedance trace to generate the Gaussian plot of Figure 31.3.

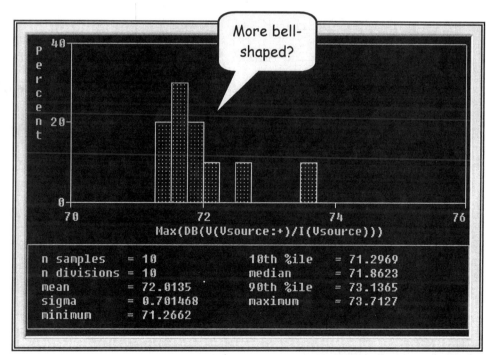

FIGURE 31.3

Histogram using Gaussian tolerance distribution

PSpice for Windows

10. Does the histogram of Figure 31.3 seem more realistic?

Yes   No

11. If you wish, repeat the histogram generation using various *seed* values.   (The seed value box within the *Monte Carlo analysis* box sets the random number generator.   That is, each seed value generates a unique random number sequence.)

12. Using the equations below, calculate the *mean* and *sigma* (standard deviation) for either of the histograms of Figures 31.2 or 31.3.   Do your answers agree with those calculated by Probe?

$$\text{mean} = \frac{1}{N} \sum_{0}^{N} X_N \qquad \text{sigma} = \frac{1}{N-1} \sqrt{\sum_{0}^{N} (X_N - \text{mean})^2}$$

Exercises

1. Assign the resistor of the RC circuit of Figure 31.4 a tolerance of 10%.   Generate a histogram that shows the probability of the risetime falling within various risetime slots.   [*Hint*: Use goal function *Risetime(1)*.]

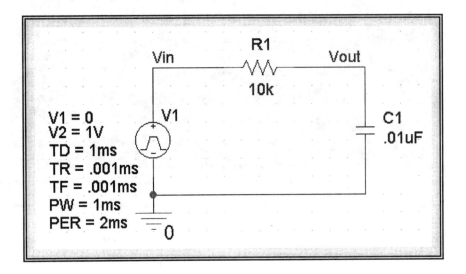

FIGURE 31.4

Risetime circuit

2. Assign all three components of a series RCL circuit a tolerance of 10%.   Generate a histogram that shows the probability of the resonant frequency falling within various frequency slots. [*Hint*: Use goal function *CenterFreq(1,db_level)*.]

3. Assign all the resistors and capacitors of the bipolar amplifier of Figure 8.1 a tolerance of 10%. Generate a histogram that shows the probability of the gain falling within various amplitude slots.

Questions and Problems

1. Whenever we perform a *Monte Carlo Analysis* and switch to *Performance Analysis*, we generate a histogram.

    True   False

2. Does performance analysis require multiple runs of a waveform?

    Yes   No

3. When using performance analysis, the X-axis is

    a. the single returned value from the goal function.
    b. the varying parameter value.
    c. the source voltage

4. When generating a histogram, the X-axis is

    a. the returned goal function values.
    b. percentage.
    c. time or frequency.

5. A goal function returns how many values when it analyzes a single waveform?

6.   When a performance analysis is run on a family of curves generated by the random tolerance variation of a Monte Carlo analysis, we generate a:

_____.

7.   Analyze goal function *centerfrequency( )* and explain how it works.

```
CenterFreq(1, db_level) = (x1+x2)/2
{
     1|Search forward level(max-db_level,p) !1
      Search forward level(max-db_level,n) !2;
}
```

8.   Why is a Gaussian-based histogram (Figure 31.3) usually more realistic than a conventional histogram (Figure 31.2)?

9.   To save time, our histograms were generated with only 25 Monte Carlo runs. What do you think either curve (uniform and Gaussian) would look like if the number of runs approached infinity?

32

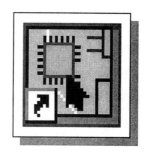

Analog Behavioral Modeling

Controlled Sources

Objectives

- *To use controlled sources to model a circuit element*
- *To add behavioral modeling to simulate more complex circuit components*

Discussion

Controlled Sources

To simplify initial circuit design and to model customized components, PSpice offers the four types of *controlled sources* listed below:

Device	Description
E	VCVS (Voltage-Controlled Voltage Source)
F	ICIS (Current-Controlled Current Source)
G	VCIS (Voltage-Controlled Current Source)
H	ICVS (Current-Controlled Voltage Source)

These devices have ideal input/output characteristics and simple transfer functions. For example, Figure 32.1 shows an *E* device (a VCVS) used as a simple voltage amplifier with a gain of 10.

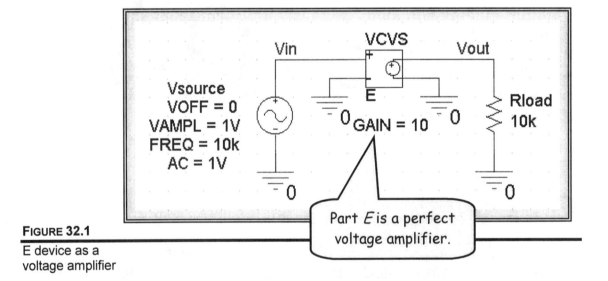

FIGURE 32.1

E device as a
voltage amplifier

By generating various time- and frequency-domain test curves, such as those in Figure 32.2, we find that the E device is a perfect voltage amplifier—with a gain of exactly 10, infinite Z_{IN}, zero Z_{OUT}, infinite bandwidth, zero harmonic distortion, and totally noiseless.

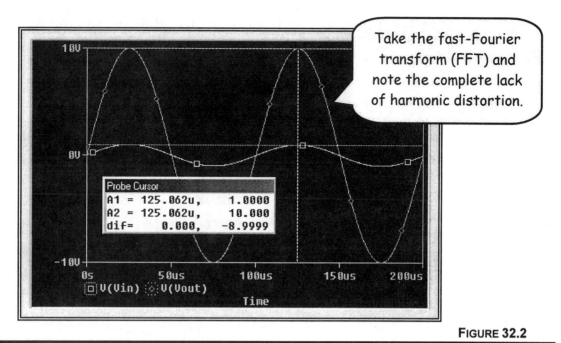

FIGURE 32.2

E device V_{IN} and V_{OUT} waveforms

Behavioral Modeling Extensions

Because the simple E, F, G, and H devices are so perfect, they are not always appropriate in a real-world environment. For this reason, PSpice also offers special extended versions of the E and G devices that provide *analog behavioral modeling*.

Analog behavioral modeling allows the designer to specify complex transfer functions that more closely represent actual circuit components and systems. For each of these special E- and G-type controlled sources, we can select from among the following mathematical relationships:

- Sum (SUM)
- Multiply (MULT)
- Table (TABLE)
- Value (VALUE)
- Frequency (FREQ)
- Laplace (LAPLACE)
- Chebyshev Filters (CHEBYSHEV)

PSpice for Windows

As an example of the practical use of these extensions, let's model a single JFET for use in a voltage amplifier. Since a JFET (when properly biased) is a VCIS with a parabolic transfer function, we select the GVALUE model. We specify the transfer function, add a load resistor and bias voltage, and generate the amplifier circuit of Figure 32.3.

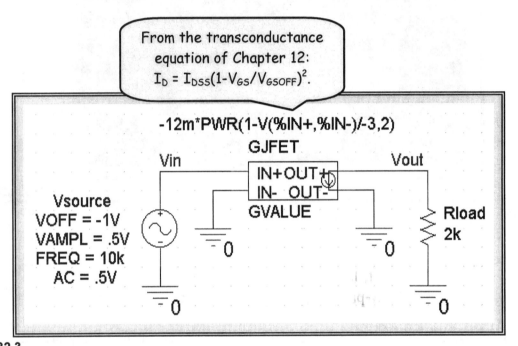

From the transconductance equation of Chapter 12:
$$I_D = I_{DSS}(1-V_{GS}/V_{GSOFF})^2.$$

-12m*PWR(1-V(%IN+,%IN-)/-3,2)

GJFET

Vin   Vout

IN+ OUT+
IN- OUT-
GVALUE

Vsource
VOFF = -1V
VAMPL = .5V
FREQ = 10k
AC = .5V

Rload
2k

FIGURE 32.3

Behavioral modeling
of a JFET amplifier

When we test the amplifier for voltage gain, it does give the expected results (Figure 32.4). (The JFET's other characteristics, such as Z_{IN}, Z_{OUT}, bandwidth, and noise, are still perfect. However, it does show the expected total harmonic distortion of approximately 6%.)

Frequency-Domain Modeling

The frequency-domain models (*FREQ* and *LAPLACE*) are considerably more complex because the output is not instantaneous with each input value, but depends on the input characteristics over time (such as its frequency).

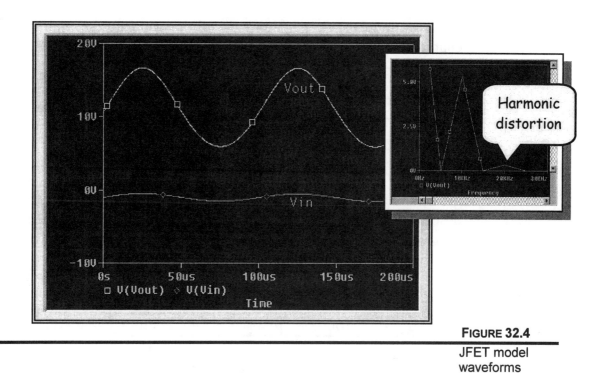

FIGURE 32.4

JFET model
waveforms

As a case in point, lets use the EFREQ device of Figure 32.5 to
model the simple low-pass filter shown in the inset.

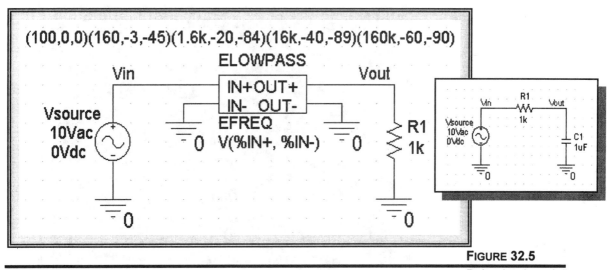

FIGURE 32.5

Behavioral modeling
of a low-pass filter

To model such a filter, we first calculate several of its characteristics:

- Fbreak = 1/(2πRC) = 159.15Hz ≈ 160Hz

- A(low frequency) = 0dB   Rolloff = 20dB/dec

- Phase = atan(X_C/R)

We then use these characteristics to form a look-up table to describe the frequency response.   These values will be used to program part *EFREQ*.

Frequency	A(dB)	Phase
100	0	0
160	–3	–45
1.6k	–20	–84
16k	–40	–89
160k	–60	–90

When we test the model (Figure 32.6), it does generate low-pass filter curves that directly reflect the table values.

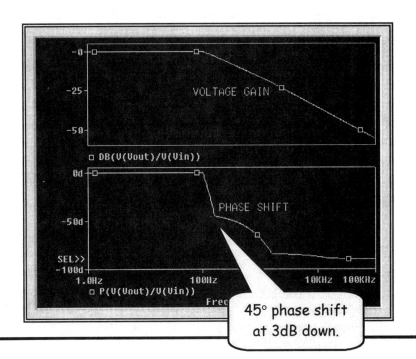

FIGURE 32.6

Filter model
waveforms

PSpice for Windows

During our *simulation practice*, we recreate the circuits of Figures 32.1, 32.3, and 32.5 and verify that they have the expected properties.

Simulation Practice

Activity *VCVS*

Activity *VCVS* will create and test the E-device voltage amplifier of Figure 32.1.

1. Create project *abm* (making sure to add library *ABM*) with schematic *VCVS*.

2. Draw the VCVS amplifier of Figure 32.1.   (The E, F, G, and H controlled-source devices are found in library *analog.slb*. **DCLICKL** on the E device to bring up the Property Editor dialog box and set the *gain* field to 10.)

3. Using conventional techniques from this and previous chapters, set the appropriate simulation profiles and generate the curves of Figure 32.2.

    By examining the results of these curves, *and others as needed*, determine each of the following:

    A =   _____   BW   =   _____

    Z_{IN} =   _____   HD   =   _____

    Z_{OUT} =   _____   ONOISE   =   _____

4. Based on the results of step 2, does the E device have perfect VCVS (voltage amplifier) characteristics?

    Yes   No

Activity *GJFET*

Activity *GJFET* will create and test the GVALUE-device of Figure 32.3, which simulates the VCIS characteristics of the JFET tested in Chapter 12.

5.   Add schematic *GJFET* to project *abm*.

6.   Draw the JFET model circuit of Figure 32.3, and set all the attributes as shown, except the transfer equation. (Part *GVALUE* is in library *abm.slb*.)

7.   To set the transfer equation of device *GJFET*, we review Chapter 12 and obtain the transconductance relationship for a JFET.

$$I_D = I_{DSS}(1 - V_{GS}/V_{GSoff})^2$$

Also from Chapter 12, we find that device J2N3819 has the characteristics: $I_{DSS} = 12mA$ and $V_{GSoff} = -3V$.

$$I_D = 12mA(1 - V_{GS}/-3)^2$$

8.   **DCLICKL** on part GJFET to bring up its *Property Editor* dialog box and enter the following in the *expr* (expression) field:

$$-12m*PWR(1 - V(\%IN+,\%IN-)/-3,2).$$

where *V(%IN+,%IN−)* refers to the voltage between the two input nodes and is, therefore, the same as V_{GS}. (The minus sign at the front of the expression is required to simulate the 180° phase shift.)

9.   Set up the system for transient and DFT Fourier analysis [10k center frequency, 3 harmonics, and output *V(Vout)*]. Run PSpice and generate the input/output curves of Figure 32.4. What is the approximate (average) gain of the amplifier?

  A = _____

10. Double the load (*Rload*) to 4k and again measure the gain. Does the gain approximately double? Does this prove that the JFET model is a voltage-controlled current source (VCIS)?

Yes   No

11. Locate the Fourier section of the output file, and record the total harmonic distortion.

HD = _____

12. Using the DC Sweep mode, generate the transconductance curve of Figure 32.7.

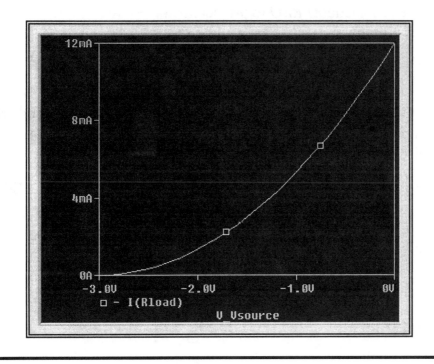

FIGURE 32.7

JFET transconductance curve

13. Based on all the results of this activity, does part GJFET successfully model part JFET of Chapter 12?

Yes   No

14. Determine the input impedance of part GJFET. Is it very high, and how does it compare to the value determined for part JFET in Chapter 12?

Activity *EFREQ*

Activity *EFREQ* will model the low-pass filter of Figure 32.5 by listing its characteristics in a table.

15. Add schematic *EFREQ* to project *abm*.

16. Draw the low-pass filter model circuit of Figure 32.5, which uses the EFREQ device (from library *abm.slb*).

17. **DCLICKL** on part *EFREQ* to bring up its Properties Editor dialog box, and enter the following data developed in the discussion into field *TABLE*.

(100,0,0)(160,–3, –45)(1.6k, –20, –84)(16k, –40, –89)(160k, –60, –90)

18. Set an AC Sweep simulation profile and generate the Bode and phase plots of Figure 32.6.   Does the device reasonably model a low-pass filter?

Advanced Activities

19. Design and test a behavioral model that converts a graph of distance versus time to a graph of velocity. (Hint: Use part *DIFFER*.)

20. Design and test a *compander* (a logarithmic amplifier found in telephone circuits to compress and decompress the waveforms). (*Hint*: Use part *LOG10*.)

Exercises

1. Design and test a behavioral model for a tank circuit of center frequency 10k and a Q of 100. (*Hint*: Use a table.)

2. By replacing the op amp with an ICVS (*H*) device, design the integrator circuit of Chapter 23 (Figure 23.3). Repeat using part *INTEG* (no feedback is needed) and compare your results.

3. Determine if the basic E device of Figure 32.1 is a differential amplifier. If so, determine its common mode rejection ratio.

4. Use controlled sources or behavioral modeling to reproduce any of the previous circuits of Volumes I and II.

Questions and Problems

1.   What are the major characteristics of a VCVS controlled source (E device)? Of an ICIS controlled source (F device)?

2.   What is the relationship between input and output of the *GMULT* device?

3.   Why should the output JFET waveform of Figure 32.4 show harmonic distortion?

4. Why is frequency-domain analysis (using an FREQ device) referred to as *non-instantaneous*?

5. Based on the E- and G-value devices studied in this chapter, would you expect the EFREQ device of Figure 32.5 to also have perfect Z_{IN} and Z_{OUT} characteristics (infinity and zero)?

Appendix A

Simulation Note Summary

Volume I

The starred notes (⋆) are repeated here in Appendix A.
For the others, you must refer directly to Volume I.

Volume II

Note	Title	Page
1.1	How do I change the X-axis variable?	9
1.2	How do I change model parameters?	12
2.1	How do I change the ambient temperature?	23
2.2	How do I set up the DC Sweep nested mode?	24
5.1	How do I uncouple plots?	66
5.2	How do I set a watch alarm?	68
10.1	How do I generate a damped sine wave?	141

Volume I Starred (✶) Simulation Notes

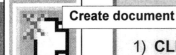

Simulation Note 1.1
How do I create a new project?

1) **CLICKL** the Create Document toolbar button (or **File**, **New**, **Project**) to bring up the New Project dialog box, fill in as shown, **OK**.

2) Select *Create a blank project* in the Create PSpice Project dialog box, **OK** to open the Capture window.

Simulation Note 1.2
How do I place parts on the schematic?

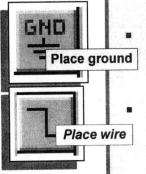

Place part

To place a part, **CLICKL** the *Place part* toolbar button to open up the Place Part dialog box. From this point on we have several options:

- The standard method is to select (**CLICKL**) the proper library, scroll to find the desired part, select (**CLICKL**) the part (to also display its symbol), **OK**, drag to desired location, (if necessary, **CLICKR**, **Rotate**, **Mirror**, or **Zoom**), **CLICKL** to place, **CLICKR**, **End Mode**.

- If we already know the part name, select the correct library (or select all the libraries) and enter it directly in the *Part* box, **OK**.

- If the part has been previously used by the project, select the *Design Cache* library, select the part, **OK**.

Place ground

- To place a ground, click the *Place Ground* toolbar button, select (**CLICKL**) *0/Source* (or just *0*), drag, **CLICKL** to place (repeat for additional grounds), **CLICKR**, **End Mode**.

Place wire

- To place a wire, click the *Place Wire* toolbar button. Click the starting point, drag to end point, **CLICKL** to anchor. Repeat for additional segments. When completely done with all segments, **CLICKR**, **End Mode**.

In all cases, be aware that multiple copies of any part or wire can be placed in succession by any number of drag, **CLICKL** cycles in sequence.

Cut

If you make a mistake, delete the component by select, **CLICKR**, **Delete**; or select, *Del* key; or, click the *Cut to Clipboard* toolbar button, which allows the component to be brought back by **Edit**, **Paste**; or click the *Undo* toolbar button to cancel the previous step. (*Redo* to bring it back.)

Undo

To delete multiple items, hold down the *Ctrl* key and select (**CLICKL**) the desired items, or **CLICKLH** and draw a box around the items.

Redo

Simulation Note 1.3
How do I "refresh" or "resize" my circuit?

- To increase/decrease circuit size in stages, **CLICKL** on the *Zoom in/out* toolbar buttons, or **CLICKR** anywhere on the screen, **Zoom in** or **Zoom out**. Position circuit with scrollbars.

- To show the entire Schematics page, **CLICKL** on the *Zoom to all* toolbar button.

- To fill the screen with the circuit or any portion of the circuit, **CLICKL** on the *Zoom to region* toolbar button, **CLICKLH** and draw a box around the selected circuit, **CLICKR**, **End Mode**.

- To move all or any portion of the circuit around the screen, **CLICKLH** and draw a box about the selected circuit, **CLICKLH** and drag to new location.

Simulation Note 1.4
How do I reposition components?

To reposition a part symbol or its associated name or value:

- Select (**CLICKL**) the desired item, followed by **CLICKLH,** and drag to desired location.

  CLICKLH = click left and hold.

- To select multiple items, **Ctrl/CLICKL** on each, or draw a box about the items. When multiple items are selected, all move as a unit.

- When repositioning labels, you may wish to click the *Snap to grid* toolbar button to turn red and disable this feature.

Snap to grid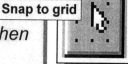

Snap to grid must be enabled (not red) when placing or moving parts!

Important!

Simulation Note 1.5
How do I change value attributes?

To change a value attribute:

   DCLICKL on the attribute to select and bring up the Display Properties dialog box. Enter the correct value in the *Value* field, select the desired display format, **OK**. If necessary, click the *Snap to grid* button and reposition the attributes.

   (For most parts, the default *Value Only* display format is satisfactory. Had we selected *Name and Value* for *V1* and *R1*, then *DC=10V* and *Value=5k* would have been displayed.)

Simulation Note 1.6
How do I set the simulation profile?

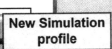

New Simulation profile

Click the *New Simulation Profile* toolbar button to bring up the New Simulation dialog box and perform the steps below:

1. In the *name* field, enter any descriptive name reflecting the type of analysis (such as *Bias Point*).

2. If available and appropriate, select a previously defined profile and inherit its characteristics. (We will seldom do this, so *none* is okay.)

3. **Create** to bring up the Simulation Settings dialog box. Under *Analysis Type*, select the desired type (such as *Bias Point*), fill in all required information, **OK**.

Simulation Note 7.4
How do I use the cursor?

To activate the cursors, **CLICKL** on the *Toggle Cursor* toolbar button. Note the appearance of the cursors and the Cursor window. If necessary, reposition the Cursor window (**CLICKLH** on the color bar and drag).

There are two cursors, A1 and A2.

- A1 is associated with closely spaced dotted lines and is controlled by the left-hand mouse button.

- A2 is associated with loosely spaced dotted lines and is controlled by the right-hand mouse button.

Cursor use is as simple as 1, 2, and 3.

1. To associate a cursor with a waveform, **CLICKL** (for cursor A1) or **CLICKR** (for cursor A2) on the appropriate color-coded legend symbols in front of the trace variables along the bottom of the graph.

2. To position either cursor, **CLICKL** or **CLICKR** at any graph location and cursor A1 or A2 moves to that horizontal location. (**CLICKLH** or **CLICKRH** to drag either cursor). The *x/y* coordinates of each cursor (as well as the difference) appear in the cursor window.

3. To fine tune A1 (or A2), use the arrow keys (or Shift + arrow keys).

To quickly move the cursor to certain key points, click any of the corresponding toolbar buttons shown below (or **Trace**, **Cursor**, select option).

View, Toolbars, Cursor to bring up.

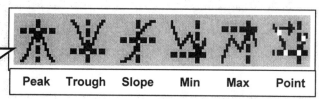

Peak   Trough   Slope   Min   Max   Point

To deactivate the cursor, **CLICKL** on *Toggle cursor* toolbar button.

Appendix B

Axis and Grid Control

Simulation Note A1.1
What control do I have over the axis settings?

To customize the axis settings for easier viewing, **DCLICKL** the corresponding axis (or **Plot**, **Axis Settings**) to bring up the *Axis Settings* dialog box. Note the first two tabs.

X-Axis

- With the Data Range box, we can let the system set the range of X-axis values (*Auto Range*), or we can click *User Defined* and fill in the desired lower and upper range.

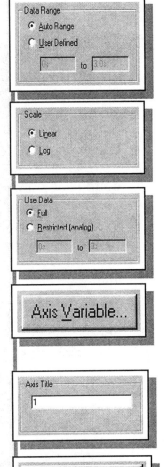

- With the Scale box, we can select *Linear* or *Log* for the X-axis scale.

- With the Use Data box, we can restrict the range of data to be displayed.

- We can change the X-axis variable by **Axis Variable** and click to select the new trace variable or expression.

Y-axis

- The *Data Range* and *Scale* options are also available for the Y-axis. In addition, we can title the Y-axis by first selecting the axis number (if more than one Y-axis), and then entering the new title in the box.

  By clicking *Save As Default*, all the new settings will automatically set for the next simulation.

Simulation Note A1.2
How do I change the grid settings?

To customize the grid settings for easier viewing, **DCLICKL** the appropriate axis (or **Plot**, **Axis Settings**) to bring up the Axis Settings dialog box.   Note the last two tabs.

X Grid

The Major section controls the solid grid lines and the placement of X-axis value; the Minor box controls the dotted lines between the major lines.   When *Automatic* is selected, the major and minor grid spacing is automatically chosen by the system.

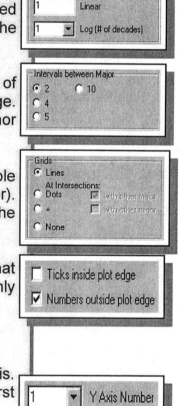

- With *automatic* disabled, the Spacing box allows us to set the major (solid) grid lines and their associated values.   For example, a linear value of 1 places the lines and values at 1, 2, 3, etc.).

    The intervals between Major box sets the number of dotted line segments that fit between each major range. For example, selecting 2 intervals places a single minor dotted line between each major solid line.)

- With the Grids box we can enable (**Lines**) or disable (**None**) the display of grid lines (both major and minor). We can also specify that dots or crosses appear at the intersection of major and minor lines.

- Finally, we can place the *ticks* (the small markers that appear along the axis) inside or outside the plot. (Only the major grids have associated numbers.)

Y-axis

- The Y-axis grid has the same controls as the X-axis. Just remember, if there is more than one Y-axis, first select the axis number.

Appendix C

Specification Sheets

D1N750

ZENER DIODE

Rating	Symbol	Value	Unit
DC Power Dissipation at TA <= 50°C Derate > 50°C	PD	500 3.3	mW mW/°C
Operating and Storage Junction Temperature Range	TJ, Tstj	-65 to +200	°C

Type Number	Nominal Zener Voltage Vz at IZT Volts	Test Current IZT mA	Maximum Zener Z Zzt at Izt Ohms	Maximum Zener Current IZM mA	TA = 25°C IR at VR = 1V uA	TA = 150°C Ir at VR = 1V uA
D1N750	4.7	20	19	75 95	2	30

2N3904/3906
NPN/PNP SILICON SWITCHING AND AMPLIFIER TRANSISTORS

Rating	Symbol	Value	Unit
Collector-Base Voltage	VCB	60	Vdc
Collector-Emitter Voltage	VCEO	40	Vdc
Emitter-Base Voltage	VEB	6.0	Vdc
Collector current	IC	200	mAdc
Total Power Dissipation at TA = 60°C	PD	250	mW
Total Power Dissipation at TA = 25°C	PD	350	mW
Derate above 25°C		2.8	mW/°C
Total Power Dissipation at TC = 25°C	PD	1.0	mW
Derate above 25°C	PD	8.0	mW/°C
Junction Operating Temperature	TJ	150	°C
Storage Temperature Range	Tstg	-55 to +150	°C
Characteristic	**Symbol**	**Max**	**Unit**
Thermal Resistance, Junction to Ambient	R0jA	357	°C/W
Thermal Resistance, Junction to Case	R0jC	125	°C/W

Characteristic	Symbol	Min	Max	Unit
Collector-Base Breakdown Voltage (IC = 10uAdc, IE = 0)	BVcbo	60		Vdc
Collector-Emitter Breakdown Voltage (IC = 1mAdc, IB = 0)	BVceo	60		Vdc
Emitter-Base Breakdown Voltage (IE = 10uAdc, IC = 0)	BVebo	6.0		Vdc
Collector Cutoff Current (VCE = 30 Vdc, VEB(off) = 3.0 Vdc)	Icex		50	nAdc
Base Cutoff Current (VCE = 30 Vdc, VEB(off) = 3.0 Vdc)	Ibl		50	nAdc
DC Current Gain (IC = 0.1 mAdc, VCE = 1.0 Vdc) (IC = 1.0 mAdc, VCE = 1.0 Vdc) (IC = 10 mAdc, VCE = 1.0 Vdc) (IC = 50 mAdc, VCE = 1.0 Vdc) (IC = 100 mAdc, VCE = 1.0 Vdc)	Hfe	40 70 100 60 30	300	
Collector-Emitter Saturation Voltage (IC = 10 mAdc, IB = 1.0 mAdc) (IC = 50 mAdc, IB = 5.0 mAdc)	VCE(sat)		.2 .3	Vdc
Base-Emitter Saturation Voltage (IC = 10 mAdc, IB = 1.0 mAdc) (IC = 50 mAdc, IB = 5.0 mAdc)	VBE(sat)	.65	.85 .95	Vdc

2N5484-5486 (Similar to 2N3819)
JFET
Maximum Ratings

Rating	Symbol	Value	Units
Drain-Gate Voltage	VDG	25	Vdc
Reverse Gate/Source Voltage	VGSR	25	Vdc
Drain Current	ID	30	mAdc
Forward Gate Current	IG(f)	10	mAdc
Total Device Dissipation at TC = 25°C Derate above 25°C	PD	310 2.82	mW mW/°C
Operating and Storage Junction Temperature Range	Tj, Tstg	−65 to +150	°C

ELECTRICAL CHARACTERISTICS

Characteristic	Symbol	Min	Typ	Max	Unit
Gate/Source Breakdown Voltage (IG = −1.0uAdc, VDS = 0)	V(BR)GSS	−25			Vdc
Gate Reverse Current (VGS = −20 Vdc, VDS = 0)	IGSS			−1.0	μAdc
Gate/Source Cutoff Voltage (VDS = 15Vdc, ID = 10nAdc)	VGS(off)	−.5		−4.0	Vdc
Zero-Gate-Voltage Drain Current (VDS = 15Vdc, ID = 10nAdc)	IDSS	4.0		10	mAdc
Forward Transfer Admittance (VDS = 15Vdc, VGS = 0, f = 1kHz)	Yfs	3500		7000	μmhos
Input Admittance (VDC = 15Vdc, VGS = 0, f = 100 MHz)	Re(Yis)			100	μmhos
Output Admittance (VDS = 15Vdc, VGS = 0, f = 1.0MHz)	Yos			60	μmhos
Output Conductance (VDS = 15Vdc, VGS = 0, f = 100MHz)	Re(Yos)			75	μmhos
Forward Transconductance (VDS = 15Vdc, VGS = 0, f = 100MHz)	Re(Yfs)	3000			μmhos

LM741
OPERATIONAL AMPLIFIER

Parameter	Conditions	Value			Units
		Min	Typ	Max	
Input Offset Voltage	TA = 25°C		1.0	5.0	mV
Input Offset Current	TA = 25°C		20	200	nA
Input Bias Current	TA = 25°C		80	500	nA
Input Resistance	TA = 15°C, VS = +-20V	3	2.0		MΩ
Large-Signal Voltage Gain	TA = 25°C, VS = +-15V	50	200		V/mV
Output Short-Circuit Current	TA = 25°C		25		mA
Common-Mode Rejection Ratio		70	90		DB
Bandwidth	TA = 25°C	437	1.5		MHz
Slew Rate	TA = 25°C, Unity Gain		.5		V/us
Supply Current	TA = 25°C		1.7	2.8	mA
Power Consumption	TA = 25°C, VS = $\pm$15V		60	100	mW

Index